Praise for *Hunger and Hope*

"[I] enjoyed the book immensely. It is noteworthy for the clarity with which it explains complex development concepts in nontechnical terms using real-world examples from both high- and low-income countries in different regions of the world."

—Robert Thompson,
Johns Hopkins University,
School of Advanced International Studies and
Senior Fellow, Chicago Council on Global Affairs

". . . readable, thought provoking, and valuable for those who want to understand the plight of rural poor in developing countries. An accessible introduction to the challenges and underpinnings of agricultural development practice, drawing thoughtful linkages between stirring human stories and lessons from professional experiences, it is a wonderful book to have on either a personal or course reading list. For those of us in the field for a while, it provides a touching reminder of why we became involved in development in the first place."

—Greg Traxler,
Senior Program Officer for Agricultural Development,
The Bill and Melinda Gates Foundation

". . . a delightful read with many insights regarding people and how economic development can work. I highly recommend it to both high school and college students interested in world development and food problems, including my own grandkids . . . a must read for teachers who want to interest their students in development and food security issues."

—K. William Easter,
Professor of Applied Economics (retired),
University of Minnesota

"Exceptionally engaging . . . could almost be recommended for its low-key humor alone. What makes it a must-read is the deft use of personal observations and worldwide experience spliced with scholarly insights that challenge the inevitability of ongoing poverty and hunger and that build hope."

—Lars Brink,
former Senior Advisor,
Agriculture and Agri-Food Canada

". . . highly enjoyable . . . part memoir and part develo--- book offers deep insights into the processes sh ⟩vides clear and detailed observations regardir ∶hallenges of the rural poor. Genuine respect for nic ladder resonates on every page, and abidi⟩ ₂ader to new levels of hope for the path ahead."

D1417165

⌐erald Shively,
Professor of Agricultural Economics,
Purdue University

"The author conveys the complexity of poverty and, as part of that, [presents] the poor as decent human beings trying to make the best of a difficult life, including a few pleasures here and there, as he takes us down the road to explaining how poverty is reduced. He shows the tremendous progress being made as well as the immensity of the remaining tasks. That [progress] gives us hope for the future. He guides us to understanding complex problems and programs by interspersing fascinating stories of real people with clearly stated facts and relationships. A combination of extensive grass roots experience and knowledge of theory, data, and policy are the unique bases for this valuable book. Read and enjoy the experience and the knowledge."

—John W. Mellor,
former Director General,
International Food Policy Research Institute

"George Norton is an internationally acclaimed scholar of the economics of agricultural development. The hallmark of his work is intensive engagement with farmers around the world to identify their aspirations, the constraints that limit their productivity, the risks that imperil their lives and livelihoods, and the opportunities they might seize to improve prospects for themselves, their families and their neighbors. This is a deeply interdisciplinary endeavor, which George reports with remarkable grace and good humor. His is a touchingly personal story, underscoring that his is more than a mere intellectual adventure or a moral crusade to help lift billions of people from unnecessary suffering. The story is at once touching and enlightening, about science as much as about human relationships. I will have my own children and students read it and learn from George's years of wisdom accumulated around the globe."

—Chris Barrett,
Professor and Director of the Dyson School
of Applied Economics and Management,
Cornell University

"A rare glimpse at development economics in the field, with the humor, frustrations, and heartbreak that come with it. The book is rich with real-world illustrations of key development concepts."

—Ed Taylor,
Professor of Agricultural and Resource Economics,
University of California, Davis

"I know of no other book like this one. It tells stories of the author's experiences growing up on a dairy farm in New York, as a Peace Corps volunteer, assisting with a tribal farm in the US northern plains, and working as an agricultural economist around the world. It weaves these stories, beautifully, with the hard lessons of the economic and institutional changes needed to give poor people hope for a better life. Professor Norton does not romanticize poor people, nor does he idealize the insights of agricultural scientists or policy makers, but shows how farmers themselves, scientists, and others can make important contributions to the process of sustainable economic development."

—Paul Heisey,
Agricultural Economist,
US Department of Agriculture

"*Hunger and Hope* is highly readable, stimulating, and told in first-person style by a professional who has earned the respect of the development community for his scholarship and commitment. Complex agricultural development problems are explained in ways that both lay and professional reader can understand, with sensitivity to cultural foundations and the reality of impoverished families. Parallels are drawn with the author's own farm experience early in life, bringing new understanding to the processes of development, and insight into topics that are too frequently overlooked or treated in more superficial ways. The book provides valuable insights to government officials, students, and other citizens who seek a better understanding of their government's role in international development, or who are concerned about the work of their church group or their favorite nonprofit organization in addressing global hunger issues. This book reminds us that there is no panacea for the challenges of feeding a hungry world, and that patience and in-depth work with very diverse farming conditions is required."

—Brady Deaton,
Professor of Agricultural Economics
and Chancellor Emeritus,
University of Missouri

Hunger and Hope

Hunger and Hope

Escaping Poverty and
Achieving Food Security in
Developing Countries

George W. Norton
Virginia Polytechnic Institute and State University

WAVELAND
PRESS, INC.
Long Grove, Illinois

For information about this book, contact:
Waveland Press, Inc.
4180 IL Route 83, Suite 101
Long Grove, IL 60047-9580
(847) 634-0081
info@waveland.com
www.waveland.com

Cover photo: Women in Bangladesh, courtesy of Sally Miller.
Photo credits: All photos by George W. Norton, except where indicated otherwise.

Contents

The Stories: Part II
Winning the World Food Fight   81

Preface

Many of the poor in the world earn their living on farms. To understand and reduce poverty and hunger, it helps to visualize conditions, hopes, and successes of farm families. *Hunger and Hope* presents real-life stories of the rural poor and lessons from field-level observations. The stories and lessons draw on my experiences as a Peace Corps volunteer working with farmers in Colombia and as an economist involved in assessing agricultural development interventions around the world over many years. The book provides a personal perspective on why some countries succeed in improving food security and others continue to struggle. It highlights the importance of institutions, the progress that has been made in reducing poverty and hunger, and the need to do more.

The prologue provides background to set the stage for the stories that follow. Stories in part I focus on poverty, hunger, health, population, environment, farming, and migration. Stories in part II suggest solutions to spur agricultural development and address the issue of violence that often constrains development. Discussion questions and suggested readings are included to facilitate study of the topics covered in the book.

All of the stories describe events that actually occurred—some of them recently and others years ago. Everyone mentioned in this book is real, although in a few cases their names have been altered to protect their privacy. I thank everyone who appears in the book, even those whose dogs bit me, for teaching me lessons in life.

Prologue
A Funny Thing Happened
on the Way to the Farm

Imelda Ospina and some of her children
with New Zealand white rabbits.

The Cow

Just after eight o'clock, I eased my motorbike into the front yard of a farmhouse on the road to San Daniel, Colombia. *"Buenas días Don Jorge,"* a woman's voice called from the smoky kitchen. *"Cómo está?"* Quickly surrounded by three small, wide-eyed children, I replied that I was fine, and after further pleasantries, asked if I could leave my bike and helmet on her porch during the day. I was headed up the mountain to Aguabonita to help Angela and Juan build a vegetable garden. Maria said, *"Claro que si"* (of course), and offered me a bowl of hot chocolate. I told her it was not necessary, but she insisted. As I drank, the children stared as if they had never seen anyone as strange as this light-faced gringo with curly hair, long sideburns, and faded jeans. After finishing, I thanked Maria, grabbed my backpack, and, as I left the yard, suggested that her kids not touch the bike until the motor cooled down.

I then began a two-hour trek up a narrow mountain trail, just above the coffee zone. Dogs barked and chickens clucked as I angled along the winding path past hard-packed dooryards of dwellings made of adobe and wood. The sun's heat slowly melted away a dense fog that shrouded the mountains, revealing a patchwork of pastures and crops from the small semisubsistence farms that dotted the hillsides.

When I reached my destination, I was surprised to see Angela slouched on the front step of her two-room shack, sobbing softly. I asked what was wrong. She slowly rose to her feet, wiped her tears, and led me around behind the house. She looked up at me with a pained expression and pointed to a black-and-white carcass lying on the ground. She said that her husband Juan had gone to an auction the day before in the nearby town of Pensilvania.[1] He had invested all of the family's meager nest egg in a big, beautiful—as cows go—Holstein. Shortly before I arrived, the bovine was blissfully chewing her cud high up in the pasture that rose sharply to the sky behind the house, when she took a misstep, lost her balance, and fell right off the mountain. The cow literally tumbled down the steep slope, landed hard on the back doorstep of the house, and expired.

For those unfamiliar with Holsteins, they are the large black-and-white cows you may have seen along the road, the most common and most productive milk breed in the United States. The Holstein is also a rather delicate creature, despite its size, and has been bred to produce a prodigious quantity of milk when pampered. The cow is well-suited for the cool, flat savanna near the capital city of Bogotá.

The typical cow in the Colombian mountains is also black and white. But it is short and has goat-like qualities that help it cling to mountainsides and consume poor quality food and water with impunity. It may not produce much milk, but it survives.

Cattle are bought and sold once a month at an auction in Pensilvania. About once a year, a dealer brings in a Holstein from Bogotá, fully aware of the old adage that a sucker is born every minute. Indeed he is—or at least once a year. The farmer sees that big, beautiful, black-and-white cow, and his natural tendency for caution is overwhelmed by the thought that bigger must be better. He buys the beast and proudly leads her home. He dreams the cow will provide ample milk for his children, as well as a surplus his wife can convert into cheese to sell in the market. In the mountains, this cow lasts as long as a tent in a tornado.

I expressed my condolences to Angela, feeling sadness and empathy for her loss. Staring at her cow, I recalled a morning when I stood in the barn doorway of the farm where I grew up and called our cows to be milked. They all raced in from the pasture in their normal haste, eager for breakfast and milking, all except one. That cow staggered through the door like a drunk, keeled over, and dropped hard to the floor. My father and I rushed to her side, but it was too late. She had sliced a major vein on the side of her udder during the night, perhaps on a barbed wire fence, and had bled to death.

Standing there with Angela, I remembered the helpless feeling when our family lost a valuable animal, and the scramble that ensued once we decided to salvage the meat. Angela and her family faced a more disastrous situation. Not only did they lose their *only* cow, but they had no freezer to store meat from a butchered animal. They would have to sell or give away what meat they could and absorb a loss they could ill afford. They still had their maize and two pigs, but the loss of the cow was devastating to their family finances.

When farmers such as Angela and Juan want to improve their situations, they must take some risks. But living close to the margin of survival, they also tend to be cautious. Farmers in developing countries need improved agricultural technologies that are appropriate for local conditions, be they animal breeds, crop practices and varieties, or farm implements. But they also need basic protection from undue risks and scams.

When societies are locally oriented, information is available to all. Sharing among relatives and friends reduces risk, and efforts by some to take advantage of others are often constrained by social and cultural norms. As economic activities expand and become less personal and more complex, informal customs no longer suffice. Political and economic institutions—the rules of society—must be formalized to improve information flows, provide incentives and safety nets, and constrain malicious behaviors. Otherwise, farmers may hesitate to try new practices, despite dreams of bettering their situations, for if they do innovate, they might suffer significant losses. Formal institutions can work for or against farmers and the poor. The design of those institutions, good or bad, goes a long way toward explaining why some countries climb up the economic ladder, while others do not—or fall back down.

Fortunately, many countries have improved their political and economic institutions, making the institutions more inclusive, providing broad-based incentives for society to invest and innovate. In these countries, farmers and others have gained enough trust in the economic and political rules that they have been willing to take risks to raise their productivity and incomes. As a result, significant strides have been made over the past half century in the global battle against poverty and hunger. Despite a doubling of the world's population during that period, fewer people are poor, famines are less frequent, and child mortality has gradually declined. For many people in developing countries, life is better than it was—and for most, offers hope.

Unfortunately, progress is uneven, and the global decline in poverty is slow, hindering efforts to improve the health and nutrition of the poor. Life expectancy exceeds 70 years in developed countries, yet fails to reach 50 in parts of the developing world. The biggest concern is for the poorest of the poor, the almost one billion people who struggle to survive on less than a dollar a day. They endure hunger, illness, and severe deprivation. Many are housed in mere huts with no latrines. Poverty stinks, figuratively and literally. The residents notice the stench, but they have little choice.

Why do the poorest of the poor in developing countries have so few choices? Why can't they just, as my grandfather once asked, "pull themselves up by their bootstraps?" The simple answer is: no boots. However, most have pride, and they strive mightily to pull themselves up by the straps of their flip-flops, cheap shoes, or bare toes. They do not wait for or expect handouts. They work hard and are affected by a combination of conditions derived from their nation's history, climate and resources, available technologies, interactions with other nations and people, and, most important, political and economic institutions. Many have lost more than their fair share of life's lotteries, but most have not given up. Some may simply accept their lot, but even they have hope that their children will do better.

Roughly 70 percent of extreme poverty globally is found in rural areas. Most of the rural poor in developing countries farm or otherwise depend on agriculture, and the poorest 20 percent of all people spend about two-thirds of their income on food. To understand why people remain poor and hungry, we must consider the conditions affecting agriculture. Those complex conditions include technologies, resources, and the institutions that influence how the economy works, but food and income problems also have a key human dimension that can be described and at least partly understood.

When I was a child, my mother would say, "Eat your peas. Remember all the starving children in China." At the time, I could not understand how eating a few peas could possibly help some poor kid on the other side of the world. When I grew older I realized I had missed her message: Be thankful for the food you have—not everyone is so lucky.

Fortunately, fewer children are hungry in China, Colombia, and other countries these days. Significant progress has been made, and lessons have been learned. We know that behavioral incentives for farmers, suppliers, marketing firms, and public officials are critical for economic growth—and that poor countries need growth for sustainable poverty reduction. But we have also learned that providing incentives for growth and then simply waiting for benefits to trickle down to those at the bottom will mean misery for many unless the political system is inclusive and safety nets are in place to protect the weak. There is always the concern that a small group may monopolize the benefits, leaving poverty for the rest of the people. A country may grow for a while with economic gains highly concentrated in the hands of a few, but eventually conflict will result and the economy will decline.[2] Balance is necessary between policies and institutions that provide incentives for growth, and those that build community and minimize risks to the most vulnerable. This book contains examples and stories that suggest what it takes to achieve that balance.

Angela and Juan were able to salvage some of the meat from their deceased cow, and I eventually built a garden with them. They continued to work hard to improve their farm—what else could they do?—and raised a large family. I lost track of them after the Peace Corps, but today most of their children are likely to be found off the farm and earning more than their parents, the implications of which are addressed below.

Farming Transition

It was a low-income country in which a few families owned large estates, but many lived in single-room mud houses with dirt floors and no chimneys. People slept on straw for beds, along with chickens and pigs. Infant mortality was high, and life expectancy was about 40 years. Only a quarter of the population could read and write. Most of the land was owned by absentee landlords, with local farmers renting tracts smaller than 10 acres. Farmers were highly dependent on a single crop for most of their food, and suffered under the oppressive policies of a nearby colonial power.[3] Unemployment was high, and many people left the country for the United States. Which country was this? It was Ireland in the early nineteenth century.[4]

Among those struggling to make a living was a young man named Patrick Naughton, who worked as a gardener on an estate in West Meath. Patrick loved to till soil and tend fruits and vegetables. He married Mary Mooney, and they soon produced a son. They were happy, but life was hard. They yearned for greater opportunities and decided to make the arduous six-week voyage across the ocean to America.

Shortly after arriving in New York, Patrick and Mary learned of land for sale upstate. They traveled north, purchased a few acres, and carved a small farm out of the rocky, hilly forest on the southern edge of the Tug Hill plateau, just below the Adirondack mountains. Patrick and Mary raised 11 children on that farm, seven of whom survived to adulthood and one of whom shared his father's first name.

The Naughton farm was too small to support the children after they grew to be adults. Young Patrick packed up and left for New York City, where the number of manufacturing and service jobs was growing rapidly.[5] He soon found employment driving a horse-drawn street car. But farming was in his blood and after a short time he returned to central New York to work as a hired hand on a local farm.

There he met a young milkmaid named Sophia Will. They married and rented a small farm "on shares," which means that they paid the rent with an agreed upon percentage of the harvest. Such "share cropping" was a common practice in the United States in the late nineteenth century, and still is in many developing countries today. Sophia's parents owned a farm in a nearby area that was aptly named Hillsboro. The farm produced several crops, dairy products, pigs, geese, maple syrup, and honey. Patrick and Sophia worked, saved their money, and in 1882 purchased a small stony patch of rolling land adjacent to Sophia's parents. They built a farm and a family of their own.[6] Building a farm entailed, among other things, constructing a house, a barn, and stone wall fences from rocks picked off the fields. Sophia helped her husband on the farm in addition to doing housework. She milked cows, fed animals, and spun, knit, and wove clothes, blankets, and sheets. Patrick died young, but their twin sons Irwin and Ernest followed in his footsteps, expanding the farm over time through the purchase of additional stony and hilly patches of land from relatives and neighbors.

In the early twentieth century, agricultural prices rose and the farm prospered, as much as a small farm could at the time. The family grew a few acres of vegetables to sell in addition to the milk they delivered to a local cheese factory. They also raised animals and crops for home consumption. Nationally, high-quality agricultural land had become scarce, limiting growth in production, while population and incomes grew rapidly, creating increased demand for food.

Irwin went off to college, and after graduation took a job in another state with Western Electric Company. He came back and married a local school teacher named Mary Clarke, but soon left again. His brother Ernie stayed, ran the farm, and took care of their mother.

The decade of the 1920s was unkind to US agriculture. Mechanization raised productivity modestly, but foreign demand for farm products, which had risen during World War I, declined, along with prices. Irwin and Mary sent money home to Ernie to help support the family farm. But food prices declined further in the 1930s, and with the nation in the grip of the Great Depression, Irwin was given an "early retirement" by his employer. With his wife and four children, he returned home to farm full time with Ernie. Throughout the United States, job scarcity in cities led underemployed labor to remain on or return to farms.

World War II brought increased prices and full employment, but also drafted two of Irwin and Mary's sons to fight. After the war, one son went to work for the railroad, the other, Elmer, came home to farm with his parents and Ernie. Elmer married Janet Williamson and they soon contributed three sons to the baby boom generation. One was me.

Growing up on that farm in the 1950s and 60s, I experienced firsthand the rapid changes that swept across US agriculture, revolutionizing the way we obtain food and sustain, or not, our rural communities. In the 1950s, our farm was diverse, much like many other farms in central New York at that time. We had two dozen cows, some calves and heifers (adolescent cows), chickens, pigs, a steer, two work horses, a tractor, a giant garden, 12 acres of corn and five of oats, 40 acres of hay and 50 of pasture, a small apple orchard, blueberry and blackberry patches, a large woodlot, and bees. We sold milk, bull calves, old cows, pigs, eggs, oats, maple syrup, logs, cider, and honey. We produced and processed much of what we ate—in popular parlance we produced and consumed "local foods." The day began with the morning milking about 6:00 AM and ended about 8:00 PM after evening chores, or even later during haying season. The farm supported eight of us, including my parents, two brothers, grandparents, and great uncle Ernie. We worked hard and had little money but were not poor.

By the mid-1950s, a second tractor had replaced the horses. Our labor force would expand in late winter when elderly relatives helped gather sap to make maple syrup, again in the spring when they helped to pick stones off fields before planting, and again in the summer when they helped with haying. These folks were mostly gone by the 1960s, and were succeeded by neighborhood youth whose pay depended on how hard they worked and how much they ate. Urban cousins who (for the experience) were "farmed out" to us in July and August, also contributed to our workforce. My aging grandfather worked in the garden, while Uncle Ernie tended his bees. My mother kept the farm accounts and did the cooking, canning, freezing, and other housework.

In spring and summer, we tuned the radio to Yankee games while preparing fields for planting, putting in hay, and cutting corn, and we played softball in the evenings after milking. As we worked and played, several economic forces were underway that eventually transformed the structure of US agriculture and the Norton farm. Those forces had begun more than a century earlier, but accelerated in the 1950s and 1960s as farms grew in size and specialization and farmers shrank in number and percentage of the overall US labor force.[7]

When the economy of a country first begins to develop, as in the United States two centuries ago and in some developing countries today, most of its people are farmers (because everyone has to eat) but production per farmer is relatively low. As agriculture develops and farmers become more productive, each farmer can feed an increasing number of people. As incomes rise due to productivity growth in agriculture and in the rest of the economy, demand gradually shifts from predominantly food to a greater amount of nonfood items, because people with money want to consume more than food, and a person can only eat so much. If agricultural productivity improves and the food supply grows, food prices will fall, unless the population is rising rapidly or a foreign market is found for the additional food. Declining prices create incentives for workers to leave farming for other occupations where they can earn more income.

Before economic development occurs, wages are low throughout the economy. Labor is abundant but capital—equipment and facilities—is scarce, so farms tend to be small. As the economy grows, wages for hired labor rise and the primary way farmers can increase their incomes is to raise their productivity through new technologies, improved management, specialization, or bigger farms. As wages grow with development, capital tends to become cheaper relative to labor, which also creates pressure to increase farm size. Farm policies can accelerate or impede this agricultural transformation to larger farms, but if a country succeeds in its development, a significant structural change is inevitable in which the farm sector declines relative to the rest of the economy.[8]

One policy that increased productivity and accelerated that change in the United States was public support for agricultural research and for education of farmers and their children. The support began in the late 1800s, and by the end of World War II it began to result in significant gains in agricultural productivity.[9] Farm production grew rapidly, and much of the large farm workforce was no longer needed. Driven by productivity growth, food prices declined despite a mass exodus of farmers.

The Norton farm felt these forces and fought hard to adjust. A new dairy barn was built in the 1950s and expanded in the 1960s. Improved corn varieties were grown, and herbicides replaced mechanical cultivation for weed control. By 1965, the chickens and pigs were history, and by 1970 the steer was, too. We could buy our eggs and meat cheaper than we could produce them. The dairy herd, however, had doubled in size. Artificial insemination from genetically superior bloodlines replaced the bull in the pasture. Mechanization increased in the milking process, and a large refrigerated tank replaced the milk cans we had used to store milk and haul it to town each morning. A tank truck stopped daily to collect the milk from our farm and transport it directly to New York City. Our milk production grew, but the milk price trended downward. We survived because our cost dropped per unit of milk produced.

By the 1980s, my brother Charles and his family were running the farm. With limited opportunity to buy additional stony patches of land in Hillsboro, they innovated, cut costs further with an improved grazing system, and supplemented their income through off-farm work. They benefited from the free labor of my brother Richard, a local school teacher. But profits were tight. By 2000, their kids had left the farm for college and higher paying jobs and the dairy herd was sold. Part of the cropland was lent to a cousin who expanded his own dairy operation. Charles and his family supplemented their off-farm income with only a small amount of farm income, primarily from hay. They still produced a few gallons of maple syrup, fruits, and vegetables for home consumption.

The transition experienced on our family farm mirrored that of countless others across the country: farmers specialized and continued to innovate and expand, left farming, or became part-timers, seeking niche or local food markets. A small number of megafarms produce most farm products in the United States today. These farms coexist with a much larger number of small operations that survive mostly on off-farm income, as our farm does now. But food is still relatively cheap, despite recent competition with biofuels, and farm households are wealthier than in the past, due largely to the income earned off the farm.

Food markets have evolved in recent years, with locally produced food, organic foods, and agro-tourism sprouting and giving new life to small, part-time farms.[10] These new marketing opportunities add youth and vibrancy to agriculture and forge connections between farmers and consumers. But the markets, while growing, remain small, and many participating part-timers today are not the same families who were squeezed out of full-time farming a generation ago.

The US farm transition was painful in many ways as farmers were forced to change or leave. Hundreds of millions of people in rural areas of the developing world are in the early to middle stages of replicating that pain, even as their overall economies improve. As was the case in the United States, the transition is laced with hope for a better future, for the few who remain in farming and for those who move on to other jobs. It is easy to become sentimental as subsistence farming wanes around the world, but life on *small, full-time* farms is hard and uncertain, especially in low-income countries. Will the rains come? Will too much rain come? Will it come at the wrong time and with wind and hail? Will insects, viruses, or diseases wipe out the crops? Will the farmer be bitten, injured, or maimed by a snake, other animal, or machine? Will prices fall or trend insidiously downward? The risks are many—there are many ways for a cow to fall off the mountain—but the rewards are potentially great for countries that can use agricultural development to feed their people and free up labor and other resources to help fuel overall economic growth. Countries that cannot do this are usually destined for continued economic stagnation. With inclusive political and economic structures that spread wealth and power and create broad-based economic incentives, this farm transition almost inevitably occurs.

While our farm in Hillsboro was caught up in the changes overtaking US agriculture in the 1960s and 1970s, I moved on as well, although never completely leaving agriculture. I attended a school that opened my eyes and then a university that taught me I hadn't seen anything yet. I embarked on a journey that would take me to hundreds of other farms in dozens of countries over many years, a journey that provides much of the material for this book. Along the way, I learned to be thankful, to appreciate that the poor are often more generous than the rich, not to underestimate the power of individuals to effect change, and to value good friends and clean sheets.

No More Cranky Cows

The journey began, innocently enough, in the automotive section of K-Mart. I was in college and had gone to the store to find a gift for a blind date to a fraternity Christmas party. I was racking my brain about what to buy, when suddenly the bulb for a "blue-light special" began to flash. It gave me an idea. So, two nights later, when the beautiful blonde opened her tastefully wrapped gift at Alpha Zeta, she was greeted with laughter as she pulled out a can of De-Icer® (pronounced 'de-ice-her'). She laughed along with everyone else and kissed me. Thank you K-Mart; she was thawed.

Fortunately, the young woman had gained a tolerance for warped humor growing up on a dairy farm with six brothers. We dated for a few months, and soon were discussing a possible future together. But her upbringing had taught Marj one unassailable fact: cows get cranky if not milked twice a day. She would not be tied down by marrying a dairy farmer. She wanted to see more of the world than a cow's rear end— growing up on a farm, you tend not to romanticize it—and had decided she wanted to go to graduate school and be a professor. Reflecting on the poor jokes I'd heard from professors, I decided that maybe professing was an occupation I could handle too. So, rather than return to the farm, I stuck with the blonde.

Marj's college major was clothing and textiles, and mine was agricultural economics. With the textile and the farm economies experiencing rapid change, those seemed like good topics to profess. But Marj said she wanted to see a bit of the world before we continued our studies, and that sounded like a good idea to me too. We decided to take as many classes on international topics as we could during the remainder of our undergraduate studies to prepare ourselves for work overseas. We took a class on the economics of agricultural development and learned that the populations in developing countries were exploding while agriculture was struggling to keep up. People were dying as a result.

What could we do? Well, why not join the Peace Corps? We had heard that the Peace Corps was seeking volunteers in agriculture and nutrition. And Marj would not have to milk cows; she could work with women and use her knowledge of textiles and nutrition. I could work with farmers. Volunteers lived at a poverty level, but we could do that; after all, we'd been students. Thus began the detour that gave direction to our lives.

As luck would have it, two of our professors were asked by the Peace Corps to recruit volunteers for Colombia and to teach a seminar course on

Colombian culture.[11] The course, which had about 15 students, also featured interesting activities such as dropping the participants off without money in a January snowstorm in random towns in upstate New York. We were to survive for three days, make friends, learn as much as possible about the town, and report back to the class. Marj and I had just been married over Christmas vacation, so this would be the first adventure of our wedded life.

When we arrived in "our town," Sherburne, we walked into the local pharmacy to escape the cold and to figure out our next move. Remembering our school days, we thought, "Perhaps the local guidance counselor can help." We went to the school and explained to the counselor, Mr. Hill, why we were there. He listened to our story and said, "You look hungry. Take these coupons to the cafeteria, get something to eat, and come back and see me."

After lunch, he told us about an elderly lady on Main Street who knew everything there was to know about Sherburne. But, it was snowing hard, and he said, "You had better take my car if you are going to visit the town. Here are the keys. Come by and pick me up after you have finished your afternoon exploring, and we will go to my house for dinner. My wife and I would be happy for you to stay with us while you are here." That evening during dinner we discussed the town and our plans for Peace Corps.

The next day, we dropped Mr. Hill off at work and continued our tour of the town. We visited the elderly lady again for more details of the town's history and current events. We visited the Rogers Conservation Education Center, a large wildlife habitat on the edge of the village. At the end of our stay, as we waited to be picked up to return to our class, Marj said to me: "This town sure has nice people."

We were given other assignments in the class that helped us gain confidence in our ability to adjust and improvise. By the end of the school year, we were ready to take on Colombia. However, our plans were almost sidetracked. I won the lottery—not the e-mail kind, but the one run by Uncle Sam. I received my draft notice, and, having no desire to study tropical agriculture in Vietnam, I marched to the office that advised students about draft options. They laughed when I walked through the door, the same laugh I used to hear at the beach. It seems the military has weight restrictions, and being skinny has its rewards. Goodbye, Vietnam. Hello, Colombia.

A few days after graduation, Marj and I left New York to fly to Bogotá via San Jose, California and Miami, Florida—Peace Corps must have used my travel agent. We trained in Colombia for 90 days, a program having much in common with college life, a school field trip, and a visit to the dentist. Yes—we went to class, took field trips, and visited the dentist. We studied Spanish for half of each day, and by the end of the first week I was no longer asking the store clerk for shaving soup (*sopa*) when I wanted shaving soap (*jabon*). We were assigned to live with a host family, the Mariños—a very nice family who helped us with the language and familiarized us with local customs, like being asked to arrive at six o'clock for dinner when the host actually wanted us to arrive at eight o'clock.

Peace Corps training can be a real eye-opener. The second week we took a field trip to the town of Guateque where the water system had been damaged by a mud slide, leaving the residents without water for a month. The sheets on the straw mattress in our hotel room had not been changed for at least that long. Ripe sheets in the tropics turn strange shades of green. We slept on top in our clothes, and discovered during the night that Marj was allergic to the fleas that had taken up residence in the bed.

Another of our training activities seemed vaguely familiar: we were to find our way without money to a small town and spend a few days learning all we could about the place. With all the snooping around, no wonder so many Colombians thought that Peace Corps volunteers were CIA agents. This time they just sent the guys. I was asked to visit the tiny town of Vianí, a three-hour bus ride from Bogotá. Being a young American male in a rural Colombian town can be an interesting experience. First, groups of girls practice their English by shouting things at you like "I love you. Please marry me. What's your name?" and then giggle. And the men try to sell you on the quality of the local brothel. Of course the girls were kidding, and the men didn't push too hard once I told them I was married, although some of the men gave me that "so what?" look.

During the warm and winding journey to the town, we successfully set the national record for number of people packed in a long-distance bus, an event duly celebrated by a chain reaction of passengers losing their lunch. Overcome by sights, sounds, and smells, I soon joined the festivities and eventually staggered off the bus sick in Vianí in the late afternoon. I was immediately met by people who were so friendly I could have sworn I was in Sherburne. They told me I was the first American to visit there in several years. Apparently, they had been saving up their hospitality. They served me food morning, noon, and night, and I never lacked for a game of basketball, soccer, or ping-pong—or for someone to talk with me. Imagine if they had known I was coming!

A farmer lent me his horse to ride around town and showed me various crops and livestock common to the region. Someone took me to meet the mayor and a group of farmers who asked me to compare farms in Hillsboro to those in Vianí. When I caught the bus back to Bogotá at the end of the weekend, I thought, "Boy, this town sure has nice people." The experience gave me added confidence that I could make friends in another culture and an eagerness to move on to my Peace Corps assignment. But first we had to finish our training in Bogotá.

Bogotá was only three hours distant, but a world away from Vianí. High up on a savanna, it had a bustling downtown and some beautiful areas surrounded by dismal slums and by mountains that seemed to hold in every ounce of exhaust spewed out by colorful buses and ancient automobiles. At that time, street kids (*gamines*) slept in cardboard boxes in doorways, hitched rides on the backs of buses and cars, begged, grabbed wristwatches from motorists at stoplights, and otherwise trained for life as hoods when they grew up. "Growing up" is not the right term, as an 8-year-old *gamine* had long since left childhood and might not survive to adulthood.[12]

Public transportation seemed plentiful and cheap to an American. But pity the pedestrian who blindly stepped off the curb into the chaotic traffic, unless the person was hailing a bus, in which case the bus might cut across four lanes of traffic to haul the new passenger on board into its mass of humanity.

Perhaps the most useful thing we learned during training was that Colombians look out for their fellow man. They lightly scratch their cheeks as they approach you on the street if they see you are in danger of being accosted—sort of like a car that flashes its headlights to warn oncoming traffic about a speed trap ahead. That cultural tidbit proved useful one day when we were trying to catch a taxi in the rain. Someone walked toward us scratching his cheek as if he had fleas. Marj and I looked at each other, turned around, and saw a man closing in on us. We darted across the street and into a crowd. The man followed us, but passed on by. We saw the knife at his side as he walked away.

We were not so fortunate another day when three men jumped us with machetes and guns and took our money, camera, and watches. I managed to bargain a little, and they did let me keep my social security and draft cards—not much, but the best I could do with a knife in my back. As they ran away, Marj couldn't believe I availed myself of the opportunity to practice my Spanish by telling the guy with the gun what I thought of his mother. If my mother had been there, she would have told me to wash my mouth out with *jabon*. Fortunately, the size of the haul seemed to please them and they were anxious to flee the scene. In hindsight, the robbers did us a favor, as they armed us with a heightened awareness of our surroundings. We were never mugged again.

Near the end of training, we were told that our "site" for the next two years would be a remote mountain village named Pensilvania. We would work with the Colombian Coffee Growers Federation (Coffee Federation) in its extension service. Made sense to me; send a kid from a New York dairy farm to teach Juan Valdes to grow coffee. Maybe we could pool our talents and make café au lait. Actually, it did make sense. We were assigned to a diversification program. Marj was to teach nutrition, health, sewing, and crafts, and I was to teach people how to raise vegetables and fruits, rabbits, chickens, and fish. Marj learned much of what she needed growing up with nine siblings. And as a kid I used to garden, fish, feed chickens, and chase rabbits.

Once we knew our specific assignments, we were dispatched on field trips to observe other volunteers who worked on similar activities. I visited Larry who built home and school gardens in a small town south of Neiva. After following him around for a couple of days, it seemed to me he worked pretty hard. When I asked what he did for entertainment, that night he took me to a small corral on the edge of town where I got to observe a cockfight. It did not take long for Larry to lose a month's wages betting with the locals, who seemed to have an uncanny knack for picking the best big bird gladiators. Among the cockfight, a bullfight we attended in Bogotá, and a pig roast that Peace Corps organized for us, I suspect an animal rights activist would have had a tough time with our training program.

Radishes and Rabbits

The first trip to our site was almost enough to make us swear off transportation. The bus left Bogotá early in the morning and, as is true the world over, the station was in the very best part of town—not. It was not so much that the neighborhood was run-down, but that questionable-looking characters lurked in nearby alleys and mingled in the crowd. We held our things tightly as we squeezed through the mass of humanity to locate the blue vehicle with red stripes that was destined for Pensilvania. The bus sported a luggage rack on top that we accessed via a ladder to secure our belongings. The call to board was drowned out by the trampling sounds of *campesinos* (farmers), other folk (young and old), and occasional livestock. Everyone knew what Marj and I were soon to find out: riding Colombian buses was like playing musical chairs—with few chairs but plenty of music.

The 12-hour ride to Pensilvania took us through all four seasons. We froze as we crawled up to ten thousand feet, roasted as we zoomed down into steaming valleys, and aged a year as we learned there were no such things as straight roads, headlights, mufflers, or speed limits. People were packed in the aisles like . . . no, not like sardines, or cattle; the odor of sardines is sweeter and cattle hold their lunch better. They were packed in like people, people who swung from side to side as the bus screamed around curves. The bus stopped everywhere and anywhere in between. The driver had a doorstep-surfing sidekick (*ayudante*) who scouted fares, pumped gas, polished the tastefully decorated dashboard, and communicated with him by pounding on the side of the bus.[13] As our driver continued to down shots of *aguardiente* (the local firewater) at each and every town—with only a modest effect on his driving—we began to understand why the *ayudante* rode along hanging ten on the step.

About four hours out of Bogotá, the bus made a pit stop in Honda, a sweltering river port city deep down along the edge of the Magdalena River. While we sat there sweating, a man standing in the aisle told us that Honda was the city of bridges, and that it had experienced a golden era in the early 1900s when it was the main transportation link between Bogotá and the Caribbean coast. We scouted the local latrine, but decided to wait until the next stop as it looked like it had been well used, but untouched by a cleaning crew since the golden era. Street sellers hawked sodas and *chicharrones* (burned pieces of hog-fat-on-a-stick) and iguana kabobs (think fried gecko-on-a-stick) through the bus windows. The latter did not smell too bad, but we were glad the Mariños had packed us some sandwiches.

An hour past Honda we left the pavement, and for the next six hours we sped down a dusty gravel road as curvy as the first letter of the towns where we stopped: Maraquita, Marquetalia, Manzanares. Shortly after sunset, the road abruptly switched back to pavement. We had arrived at the south end of Pensilvania. The sound of the bus echoed off the walls of the wooden and stucco houses that lined each side of the street as we roared to our final stop in the town plaza. We stepped off the bus, and immediately were swarmed by kids offering to carry our stuff while practicing the only two words of English they knew: "Hey mister." We were tired after the long ride, but as we stood there in the town that would be our home for the next two years, we realized we really were Peace Corps volunteers.

How did the local people feel about the arrival of two American Peace Corps volunteers? Actually, they seemed extraordinarily appreciative, more than we expected. We later learned why: two years earlier a Peace Corps volunteer named Miguel—Michael Kotzian—had lived and worked in Pensilvania. He was friendly, open, and loved by many, and had lost his life in a bus accident outside of town.[14]

Pensilvania can be described as a small village perched on the side of a mountain at the end of a road overlooking a valley. Whitewashed wooden townhouses with 8,000 people were packed into an area that seemed like it couldn't possibly hold 500. Like most Colombian towns, it had a square, a cathedral, and numerous open air cafes that blasted Latin music well into the night. It reminded us of a town in an old western, except men wore more machetes than pistols and an occasional vehicle scattered the horses. Young ladies paraded, arm in arm, up and down the street to attract the attention of young males, and a bank of clouds enveloped the town at night.

The cafes were packed in the evenings and weekends. Men sat in cliques drinking *aguardiente*, warm beer, or rum and Coke. Women and children drifted in and out of the cafes, but tended to leave late in the day when things could get a little rough: a fight, a man riding his horse into the cafe, an occasional shooting. But the town was pretty tame, and the people had the same hopes, fears, and frustrations as those in small-town America, except they probably drank more and used fewer drugs.

Pensilvania seemed an unlikely name for a town in Colombia, but no more so than a nearby town called Filadelfia. I think the town names confused some of the locals too because one day a farmer asked us if the United States was closer than Filadelfia.[15]

We made friends with numerous farmers and town folk. As Peace Corps volunteers, we earned a living allowance of $75 apiece per month, which put us near the top of the income and social scale. When in town, we mostly socialized with other *ricos* (rich people): the local extension agent, the veterinarian, the man who ran the coffee cooperative, the loan officer in the local bank, and several storekeepers. When we joined the Peace Corps we expected to live poor, but the real adjustment was learning how to live as one of the richest people in town.

A popular diversion in Pensilvania was watching the town soccer team play the next town on a field carved out of the side of the mountain. Much of the time was spent retrieving the ball. One day the town witnessed the biggest event since the earthquake that leveled the church. A helicopter bearing a presidential candidate swooped down onto the soccer field. After a speech that literally promised a chicken in every pot, it swooped back up. Some years later, I ran into a Colombian official visiting the United States. He told me he had been to Pensilvania only one time, on a helicopter during a presidential campaign. I didn't ask about the chickens.

We lived at first in a room in the building owned by the Coffee Federation. Its main attraction was that it had the only hot water in town. But the room was small and stuffy as the window didn't open, so eventually we moved to more spacious quarters. Our new "townhouse" style apartment had a chicken coop underneath, vertical plank walls with see-through cracks, plenty of fresh air, and a pig pen out back with three of the owner's pigs who doubled as a garbage disposal. Unlike our first apartment, it had a kitchen where we could use a small kerosene burner for cooking.

Because we lacked a refrigerator and were far from a city, we ate whatever foods were fresh and locally available. Cows and pigs were slaughtered two days a week, and on those days we ventured out early to buy meat in the market. The sides of the animals hung from hooks, and the butcher would use a machete to hack off a hunk for us. We chose meat with flies on it because that meant it had not been sprayed with insecticide. Back in the apartment, I would trim off the fly-specked portion and throw it off our porch to vultures that waited in a tree. On days without meat, we typically consumed beans and rice, eggs and plantain, or fast food such as bananas and beer. The restaurant in town served eggs à la cockroach, so we usually chose to eat at home.

The extension service for the Coffee Federation had an office in town where we worked on Saturday, but Monday through Friday, before daybreak, we would travel by bus into the *campo* (countryside) and return after dark. Unlike the bus from Bogotá, which was the style of an American school bus, the *campo* bus began its life as a flatbed truck. A dozen benches were attached to the bed, each the width of the truck with about six inches of legroom. One side was open and passengers climbed, via a running board, into the colorfully painted transport. It reminded us of Christmas with its basic green color and red accents painted on the outside and shrines to Jesus, the Virgin Mary, and St. Christopher on the dashboard inside—the latter tastefully framed by a fringed windshield. It was best to get to the bus by 5:30 because once it was full of people, you sat in a blue rack on the roof—where the ride in early morning temperatures was invigorating to say the least, especially if it rained.

The bus would travel about an hour into the countryside on a dirt road, dropping off farmers, beer, soft drinks, and other essentials until, after crossing 13 streams, it arrived in San Daniel, which was comprised of a small group of houses and a church clinging to a ridge. The bus would make a second run late in the day. On the return trip, the bus would haul coffee, pigs, chickens, people, and bananas.

Once in the countryside we would walk to different *veredas* (rural neighborhoods). The local agricultural extension agent, Luis Carlos, had organized 11 men's groups, which he visited on a rotating basis about every two weeks to provide technical instruction. One day he would focus on coffee diseases, another on plantain pests, and another on castrating calves. Luis Carlos was an incredibly funny, hard-drinking, charming, compassionate person whom the *campesinos* loved and trusted. He introduced us to the farmers, giving us a credibility we didn't deserve.

Marj ("Doña Margarita") organized 11 women's groups in the same *veredas*, and the reception she received was phenomenal. Marj is a very friendly person and fantastic listener. The people responded to her by opening their homes and their hearts. Each morning she would visit a home to which she had been invited, and every afternoon, after being fed a banquet the family couldn't afford, the women's group would meet, often to work on clothing or textile crafts. Marj would give brief talks on topics such as nutrition, but she let the groups set their own agendas, and sometimes they would ask her to bring the veterinarian to demonstrate how to vaccinate chickens, the priest to save souls, or the dentist to pull teeth.

I visited farms to help establish vegetable gardens and advise on pest management, build rabbit hutches, show farmers how to graft fruit trees, and dig fish ponds. I also spoke to Luis Carlos's men's groups and Marj's women's groups and worked with school kids. Much of what I taught I had learned from Luis Carlos or from my many mistakes. I discovered that Colombia will never lack for insect and disease biodiversity; that rabbits die easily, but are tough to kill—they are cute suckers; that fish prefer a big water hole to a small one; and that Peace Corps is more work than dairy farming.

It took several months before most people could understand my Spanish. If I didn't know a word or its gender, I had a tendency to just fill in the blank with the first one that came to mind or with an English word I pronounced with a unique Spanish accent. I learned that tablé, deserté, and machiné are not really Spanish words, and that a cow was transgendered when I put an *el* instead of a *la* before "*vaca*." People were generally patient as I butchered their native tongue, especially if we had each had a beer. They seemed pleased that I was trying out their language, helped me with it, and pretended to understand my speech until it improved to the point they actually did.

The local farmers worked hard, and most were willing to try something new, but adequate solutions were absent for many of their problems. They needed credit, insurance, roads, high-quality seeds, new technologies, and better health care. One technology alone, a disease-resistant tomato variety, would have made an enormous difference to their diets and incomes had it existed. They loved tomatoes, but growing them was a major challenge given the prevalence of foliar diseases.

Marj and I would often visit the same *vereda* on the same day, but usually not the same farms. When I would visit a farm to help with a garden or rabbit hutch, the wife or daughters in the family would often ask when Doña Margarita was coming. I sometimes wondered if they had invited me for my agricultural assistance or to finagle a visit from my wife.

After about six months of rising early to catch the *campo* bus, doing physical labor most of the day and getting home in the middle of the evening, we began to wear out. Fortunately, Peace Corps came to our rescue and sent us a motorcycle, freeing us from the bus schedule and reducing the risk that they would lose another volunteer, as they had Miguel.

Our Honda created quite a stir. Actually, it wasn't the bike so much as the helmets and the glorious crash I had the first time I jumped on it. Surrounded by an onslaught of curious kids, I kick-started the motor and gave a quick twist to the accelerator. The motorcycle leapt forward and sped three blocks before I totally lost control and slid sideways down the road. For several days afterward, we had to sneak out of town at odd hours to avoid hoards of kids hoping to see the astronaut-looking *gringos* perform another death-defying crash. If they spotted us, they would follow us around with their mouths literally hanging open in dumbstruck amazement until we took off. The bike was a blessing though, and only occasionally broke down, although I was forced to learn more about motorcycle maintenance than I planned. I discovered that it is a lot easier to take a motor apart than to put one back together. Fortunately, we were only stranded once on the road with a repair problem.

During the time we spent in Pensilvania, we observed many hopes and joys, heart-wrenching tragedies, irritating displays of greed, and incredible acts of kindness and compassion. Examples are provided among the stories that follow. Our Peace Corps experience lasted two years, but it left us a lifetime of images and lessons.

We conceived a child while we were in Pensilvania, and two months before we left there Marj bore our son John, somewhat to the consternation of Peace Corps, but much to the joy of those with whom we lived and worked. Some people had difficulty comprehending how we managed to go so long without having a child. They almost always had a baby nine to ten months after marriage. They were greatly relieved when ours finally arrived. A few people had started a rumor that we were really brother and sister, and that I was only claiming Marj was my wife to protect her from Colombian men. After all, Marj's last name was the same as mine rather than *de* Norton, as it would be in Colombia. People who believed the rumor were shocked when Marj became pregnant. The members of Marj's women's groups were enthralled with our blue-eyed baby and showered us with gifts.

At the end of our Peace Corps tour, we were sad, but a little relieved, to leave and return home. We were ready to see our families again, but would not miss the amoebas and other parasites that so often had intruded at inconvenient times. We had made many friends, whose willingness to open their hearts and homes we will never forget. We observed hardship, but much hospitality and hope. We knew we wanted to continue to work on food and clothing problems that affected the poor. We learned that working with farmers in a developing country can be rewarding, but that improved policies, infrastructure, education, and technologies were needed to improve their lot.

Let Them Eat Grass

Tribal leaders stared at the beautiful blue sky, and then at the millions of thin wheat seedlings that seemed to shrivel before their eyes in the large field that stretched before them. "Maybe it will rain tomorrow," they thought, "saving our crop, and our investment in seed, fertilizer, fuel, and labor." But tomorrows came and went, and the sky remained clear and calm, oblivious to their need for water. The leaders decided they needed a new plan before the next cropping season, one that would better withstand weather and other risks. Otherwise, the tribe would continue to suffer economic hardship—and yet another indignity.

The tribe's economic ills and indignities traced back more than a century, to an intrusion by white settlers and to a subsequent war with settlers and soldiers in the Minnesota River valley. Prior to the war, these Sioux people had been forced to cede large tracts of land to the settlers in exchange for long-term payments.[16] The money was used to purchase food and other goods from traders. In the summer of 1862, the payments were late and they suffered severe hardship and hunger. In early August, representatives of the Sisseton and Wahpeton bands of the Sioux (Dakota) met with a federal agent and traders who gave them food on credit until the payments arrived. Farther south near Redwood Falls, representatives from other bands met with a federal agent and traders who were less sympathetic. One trader is purported to have said, "If they are hungry, let them eat grass." That comment may not have been the spark, but the extreme deprivation and humiliation experienced by the Dakota led some of them to begin attacking settlers. Brutal battles were fought all along the river until the Indians were defeated. More than 1,000 were arrested, and on December 26, 1862, 38 Dakota prisoners were hanged in Mankato in the largest mass execution in US history. A few months later, the remaining Dakota Indians were moved to reservations further west. The Sisseton and Wahpeton bands were placed on a triangular piece of land known as the Lake Traverse Reservation in the northeastern corner of South Dakota and southeastern tip of North Dakota. They were told to farm.

And they did farm, or at least the women did, until the tribe lost part of the reservation to the railroad in the 1870s and part to homesteaders in the 1890s. Worse, each family was allocated 160 acres and forbidden to sell the land, even to others in the tribe, until the 1920s. Therefore, after a couple of generations of subdividing among heirs, each small tract had dozens of owners. Even if tribal members moved away, they still owned a piece of the land, making it hard to get agreement on what to grow. What is owned by many is controlled by none, as some developing countries are discovering today. Eventually, tribal members were allowed to rent out their plots. Over time, more and more land was rented to whites, and the economic base stagnated. The United States treated Native Americans much like colonial powers had treated indigenous people in other parts of the world, and the result was similar: persistent poverty. By the early 1970s, 50 percent of the tribe was unemployed, housing was substandard, alcoholism was rampant, and drugs were becoming a serious problem.

Then someone suggested suing the federal government. When the railroad had come through, the government had paid the Indians 10 cents an acre for the right of way. The going price for land at that time was $1.10. A court agreed that the tribe had been cheated and awarded it $1 per acre on a few thousand acres, plus interest. Oh, the wonders of compound interest over 100 years. The tribe decided to divide up most of the award per capita, but used 20 percent as leverage for a loan of several million dollars. The loan was used to buy up land from members who wanted to sell, so the tribe could consolidate the land into workable farms. The tribe decided to sell or rent the consolidated lands to individuals, and to use some of it for a tribal farm. The idea was to generate income and employment and provide a demonstration farm for individual Indian farmers.

The tribe went out and bought a mass of machinery, dove into farming, and had its finances devastated by drought in 1976. Afterwards, they came to the Federal Reserve Bank in Minneapolis and asked it to find someone to help them plan their farming activities more carefully.

While in the Peace Corps, Marj and I had applied to graduate schools, eventually deciding on the University of Minnesota. It had programs in both our fields, a former professor had recommended it, and the frozen tundra didn't sound too bad after the sweltering tropics. So once back in the states, we purchased an old station wagon and drove out to the land of 10,000 lakes and the Minnesota Twins, arriving on January 1, 1974, one of the coldest days ever recorded in the Twin Cities. We stopped at a phone booth and spent our first hour in St. Paul removing the phone that had frozen to Marj's hair due to the moisture in her breath.

We located a Colombian babysitter for John—so he wouldn't forget his roots—and jumped into courses. I studied farm and development economics, and learned much about how and why agricultural development policies are screwed up—and what might be done to fix them. Marj studied textiles.

Before we knew it, classes were behind us and we were looking for dissertation topics. I had hoped to do research related to agricultural development in Colombia, but my advisor was thinking tribal farming in South Dakota. The Federal Reserve Bank had come to the university seeking an economist who knew something about farming who could help the Sisseton Wahpeton Sioux[17] with their planning. I knew nothing about Native Americans, but knew a little about economics and farming, and since South Dakota was closer than Colombia, I figured why not.

So I hopped in the old station wagon and headed west, about five hours, to Sisseton. On my first visit, I met the tribal chairman, Jerry Flute; the planning director, Ed Red Owl; and the assistant planning director, Mike Selvage. They were friendly, didn't push firewater on me as some Colombians had, and clearly wanted help with their farms. After attending a traditional *wacipi* (powwow) in which everyone danced in a customary circle, even inviting me to dance with them, I decided maybe I could give their farm plans some direction. We prepared a proposal, the Federal Reserve Bank helped us find support, and I jumped in with both feet.[18]

I rented an apartment in a partly constructed building in Sisseton and camped there for the summer with Marj, who was working on her own dissertation, and John. With the help of Ed Red Owl and others in the tribe, I gathered information on the land, water, human capital, and other tribal resources. We planned on the computer and developed a large tribal farm, investing in irrigation equipment, beef cattle, and buffalo. We did not do too badly—considering that farming was regarded as women's work—as long as we could keep the buffalo in line. I quickly learned that two male buffalo in close proximity spar like professional wrestlers, minus the fake moves, as they compete for the minds and hearts—and other parts—of female buffalo.

The fact that the Dakota frown on bossing each other around created labor management challenges. For example, if a tribal member employed on the farm failed to show up for work, no one complained and the person would still be paid. Fortunately, Jerry Flute was up to the task of solving this and more complex problems. In fact, he was an amazing leader with a deep understanding of people and tribal culture. He would sit in Tribal Council meetings and say almost nothing, interjecting just a few words at critical moments, and leave with everything he wanted.

I spent three years working with the tribe, gaining profound respect for the ability of its members to persevere in difficult circumstances. I also learned about the lingering effects of misguided institutional changes, even a well-intentioned one such as prohibiting tribal members from selling their land for several years. It was a lesson that would repeat itself over and over in the years to come.

The Sisseton Wahpeton Oyate are proud of their heritage—many members still speak Dakota. They want to preserve the key elements of their culture while finding a way to succeed as a people in today's world. They revere buffalo, don't order each other around much, value the family clan as much or more than the tribe,[19] welcome strangers, and honor the warrior, the elder, and those who share. They have only gradually come to value higher education, in part because young Native Americans historically were sent by the federal government to boarding schools in an attempt to assimilate them into white culture. Many parents would rather see their children join the military—an honorable profession and one from which they would likely return to the reservation.

In many ways, the tribe still struggles to overcome the effects of its banishment to the reservation more than a century and a half ago. High unemployment and poverty remain serious problems. But with pride and periods of enlightened leadership, they have gradually made progress. They look to the future with hope. They now own a casino and hotel complex that provides some income to tribal members and government, although tribal finances remain precarious.

Ed Red Owl asked me to stay and work with the tribe after graduate school, but I declined. Although I had learned a lot, and there was still much to do, Marj and I hoped to teach at a university. We also had to consider something else that had intruded into our lives.

Several months after we arrived in Sisseton, Marj went back to the Twin Cities for a few days to meet with her PhD advisor. I stayed in Sisseton. Late one night the phone rang. Marj was calling from a hospital. Her legs had gone numb so she went to see a doctor, who sent her to the hospital for tests. The doctor there told her she had a "demyelinating" illness. I said I'd drive right there, but first I found a local physician in the phone book and called to ask him what demyelinating meant. He said it is a nervous system disorder, like multiple sclerosis (MS). I tried to remember what I'd heard about MS, but my knowledge was limited so I asked about her prognosis. "Well I haven't seen your wife, but we all have to die some time," he said. Although technically correct, a delay in the inevitable is generally preferred, so that was not what I had hoped to hear. I suspect the late night call might have challenged his bedside manner. As it turned out, his comment was misleading—people with MS do not live appreciably shorter lives than the rest of us. But it got me off the phone in a hurry.

Before leaving the reservation, I stopped by to tell the tribal chairman I would be away for a few days. He tried to console me by saying, "George, doctors aren't always to be believed." Nonetheless, it was a long drive back to the Twin Cities that night.

When I arrived home, Marj was still in the hospital and our son was staying with friends. Over the next few days she underwent tests to confirm her diagnosis and receive treatment. One afternoon, a doctor came into her room to discuss what is known about MS. He said that stress can exacerbate it and recommended that Marj quit graduate school and stay at home to minimize stress. He suggested she "refocus her life goals." Marj was down for a few days after that discussion—until a young resident stopped by to say he had overheard the doctor (his boss). He said, "Marj, doctors aren't always to be believed." He told her not to overdo it when it came to extreme temperatures and other stresses, but suggested she do whatever she had planned before, as much as she could, for as long as she could. With that, Marj dropped the life-has-short-changed-me sentiment, said "screw you" (figuratively speaking) to his boss, put a cot in her lab so she could rest when necessary, and finished her PhD. It reminded me how the Colombians used to call her "*muy guapa*," which translates in most places as "very pretty," but in Colombia it can mean "pretty tough."

After completing our degrees, Marj and I joined the faculty at Virginia Tech and moved to the land of the Hokies. She teaches and conducts research on textiles and apparel. I teach courses on the economics of agricultural development and assess economic impacts of technologies and policies that affect agriculture at home and abroad. Marj stays close to campus most of the time because of her MS, but my research and teaching take me to agricultural areas in developing countries about two months a year. I suppose I could be called a free-range professor, as I prefer not to spend all my time cooped up on campus. I like to wander the world in my work. The following stories reflect part of what I have observed during those wanderings over the years about poverty, hunger, disease, and other topics that can make you appreciate what you have, perhaps squirm, and occasionally chuckle.

The Stories: Part I

❖

Of Rice and Women

Farm couple weeding in Bangladesh.

1

Poverty Is Personal

Julio and Imelda Ospina lived in a tiny two-room shack on a small patch of land, high up on a ridge atop a steep mountain slope in Colombia. The walls of their house were made of weathered boards that had become blackened at the top with soot due to smoke flowing from an open stove inside. The roof consisted of sheets of corrugated tin laid over wooden poles. The climate was cool, which constrained most cropping choices. The Ospinas grew potatoes, collard greens, and onions to feed their dozen children, two of whom were twins and one of whom struggled to walk due to polio.

I first met Julio and Imelda in the Peace Corps during a meeting of farmers in their local neighborhood of San Juan to discuss the benefits of vegetable gardens and rabbits. Julio was a typical looking Colombian *campesino*, dark haired and tanned with calloused hands and feet, worn-looking pants and shirt, and ever-present felt fedora. The toughness and width of his feet were especially noticeable. I wondered if their condition reflected the fact that they were seldom constrained by footwear while he worked on the farm. Imelda was thin, wore her dark hair tied in a pony-tail, and dressed in a simple skirt, blouse, and shoes. Julio and Imelda were not yet forty, but looked older, wrinkled by life's struggles.

Julio was full of questions in the meeting. He asked about possibilities for marketing the surplus if they were to build a large garden, how to control cabbage pests, and where to obtain fresh vegetable seeds. He came up to me afterwards and asked if I would visit their farm to help plant a vegetable garden and build a rabbit hutch. I said yes, and a few days later Julio met me when I stepped off the bus in San Juan. We hiked straight up the steep pasture above the road for a half hour until we arrived at his farm, which was literally in the clouds. After a drink of hot brown sugar water, Julio, Imelda, the twins, and I began tilling the garden. Before I arrived, Julio had built a fence to keep the chickens out, and the kids had collected as much animal manure as they could find for fertilizer.

Building a garden on a mountain ridge can be a challenge. Even with five of us tilling the soil with hoes and constructing terraces, it took the morning to create about eight vegetable beds. Before planting the seeds, Imelda invited me to eat lunch with them. She asked me to sit on a small wooden bench in front of their house, and one of her daughters handed me a boiling bowl of potato soup with an *arepa* (flatbread made from corn). Their family sat in the kitchen on benches around the stove, a waist-high wooden box supported by legs and packed to the brim with soil. Steam rolled out of metal pots nestled in the coals that lay on top of the soil.

After lunch, we planted the garden. As we planted, Julio and Imelda's four-year-old daughter, Estela, mimicked my steps around the garden. In the late afternoon, I caught the bus back to Pensilvania. I returned to their farm a few weeks later, and we constructed the rabbit hutch. In a few months, their efforts paid off. Julio began to sell sacks of cabbages, and Imelda added rabbit meat, onions, and carrots to their stew. They saved a few pesos.

One day, Julio came down the mountain, took the bus to Pensilvania, and spent part of the family savings on two little piglets. It was common for a farmer in the area to raise a pig or two, which could be fed vegetable scraps, corncobs, and other waste—a sort of piggy bank that could eventually be sold or bred. Such was the porcine investment plan of Julio and Imelda.

A few weeks later, when I visited Julio and Imelda, they had sold one of the pigs, using the money to meet basic needs. I noticed, though, that they kept the second one. I asked Julio what he intended to do with it. He said he would show me and led me down the mountain to meet their neighbor, a widow with two young children. The widow lived in a shabby shack with a leaky roof, and had only an old burlap sack for a bed. In fact, one of her kids was sick and sack ridden, getting soaked whenever it rained.

Julio suggested that we help her build a rabbit hutch and a garden. He said his family would raffle off their second pig to raise money to buy materials for a new roof for her house. A few days later, he went into town and sold tickets for the piggy-ball jackpot. There were plenty of eager takers. Soon the roof was installed and the other tasks completed to repair the widow's house.

Many rural poor survive through sharing what little they have, as Julio and his family did with their destitute neighbor. When your luck is cold, a neighbor with a warm heart can certainly help. Unfortunately, many of the poorest of the poor are orphans, widows, and elderly with weak sharing networks—few people to watch their backs.

Most poor people work hard. They survive through work, caution, sharing, and hope. Economic growth and individual effort have lifted millions of Julios and Imeldas out of poverty over the past two decades. In the words of economists, markets have helped like an "invisible hand."[20] Unfortunately, millions more continue to struggle mightily to avoid being stomped by invisible feet. Those feet may be attached to people or firms with privileged access to information or unfair advantages due to their financial and political positions. They may be attached to corrupt officials, uncaring institutions, or employers who discriminate.

The poor have the same desires, willpower, and weaknesses as the rest of us. They have many talents, but often could use a helping hand, or at least a fair shake. That hand might be extended by a neighbor, but it also might come from a well-targeted policy or institutional reform, increased access to education or appropriate technology, or social safety nets such as food or financial assistance. Poverty has many faces and is certainly context-specific. The story of Julio and Imelda and the other stories that follow shine a light on a few of the faces of the poor and their coping mechanisms.

Poverty is especially prevalent in sub-Saharan Africa and South Asia. Bangladesh is one of the poorest countries in South Asia. I frequently travel there to plan and review progress on an agricultural research project related to vegetable pest management. When I go, I fly into Dhaka and take a taxi to the place where I stay. It doesn't take long to find the poor. They find me. Stuck in traffic at an intersection, first one beggar and then another begins tapping on my car window. Soon there is a chorus of light tapping by a blind beggar, a smudgy-faced child beggar, a child beggar leading a blind beggar, a woman beggar holding an emaciated baby, an armless sore-infested beggar with nose pressed against the car window—all competing candidates for least lucky in the lottery of life. Quietly pleading for *taka*, they tap softly, perhaps to avoid irritating while trying to evoke compassion.

I have found I need a strategy for responding to these beggars. Otherwise, guilt can linger long after the traffic moves on. I learned in the Peace Corps that I cannot give to them all. Often, the more I respond, the more I am swarmed by others, and the worse I feel. And I cannot rely on the strategy I sometimes use for the beggar on a Washington, DC street corner. He looks like he might be able to work, so I don't feel too bad just walking by, though a twinge of guilt lingers. But what are the alternatives for the Dhaka downtrodden?

I take my lead from the locals and keep a few coins in my pocket. If only one person is pleading, I slip a few *taka*—a few cents—into his or her hand. If there are multiple beggars, it is first-come, first-serve for a few. That works as long as I have a clear exit lane. If I am with friends and they give, I may assume they gave for me too. Occasionally someone mentions the futility of giving to street people. I acknowledge that, but have to admit, beggars make me squirm. But for the luck of the draw, that could be me. It is more than that though. It is even more than the sympathy felt for those with hunger and pain. It is the staring into faces of people who have lost their last shred of human dignity. I am often struck by the dignity of the poor, but these poorest of the poor have long since swallowed theirs in a desperate attempt to survive.

Traveling north out of Dhaka, I am headed toward an agricultural research station and pass building upon multistoried building of apparel factories along the way. Women and men, many of them teenagers, work long tedious hours in these factories, day and night, spinning, knitting, dyeing, cutting, and sewing the yarn, fabric, and garments that we wear every day. Relative to other Bangladeshis, these people have it better than some. They have a consistent source of income, albeit small, but they also risk injury or death due to unsafe working conditions in some factories. Hundreds of the more than three million apparel workers in Bangladesh have died in factory fires or other accidents in the past five years.

Continuing on, buses and rickshaws wrestle for the road. Turning left onto another street about half an hour out, I pass a scene that appears straight out of an apocalyptic flick. A vast, stark landscape is littered with tall smokestacks spewing dark, filthy dust high into the sky. The stacks sit on ovens that are used to fire rows and rows of bricks, lined up before them like soldiers on parade. Small drab buildings scattered here and there complete the surreal landscape.

A low-lying country with copious clay but few rocks or stones, Bangladesh has a huge demand for bricks and manufactures these blocks in massive quantities. Not all are destined for houses and walls. Many are broken up and used as a gravel sub-base for roads.

Piles and piles of bricks dot the landscape, several with young children or elderly folk perched on top, each with a bucket and hammer. They sit hunched over, striking the bricks, over and over, one by one. They slowly fill bucket after bucket with hand-made gravel. Breaking bricks by hand is hard, dirty work—but another survival strategy for the poor.

The poor in Bangladesh, and in many other countries, have limited access to assets such as land or livestock to raise their productivity or to cushion the effects of a crisis. Loss of assets is one factor that can push people into poverty, although destitution differs by degree.

The poor can be divided into the partly poor, the plainly poor, and the poorest of the poor. In Bangladesh and many other places, the partly poor commonly farm rented land while working at low-wage jobs (such as those in apparel factories) to diversify their sources of income. They suffer periodic food shortages and have low social standing. Torn between hope and fear, they may be one major illness or drought away from slipping into the ranks of the plainly poor.

The plainly poor are landless and have the lowest paying jobs—think breaking bricks to survive. They eat and dress poorly and cannot afford health care or education for their kids. It is best to be the firstborn child, as food dwindles with each new arrival. Families live in sorry shacks of mud and thatch or tin and plastic, and in urban areas sometimes beg or steal to survive. Parents may apologize to their kids for not having enough to eat, but still hold a glimmer of hope.

The poorest of the poor are truly destitute, with so few assets they are basically shackless. Living quarters may amount to a lean-to constructed of old plastic or netting, set against a wall along a public street. These families are usually headed by women or elderly men, many of them disabled. They have no access to resources from family or friends. At times looked down upon by even the plainly poor, who may feel a sense of upward mobility in comparison, they mostly survive by begging. They face constant risk of starvation or succumbing to illnesses the well-nourished would survive. Kids, if they survive, expect no apology for having too little to eat, and actively seek money and food wherever they can find them. Keeping hope alive is a major challenge due to the lack of food, health, and control over their lives.

The poor anywhere sometimes take extreme measures to survive. I was once reminded of that in Honduras, where I had gone as part of a team to help plan a new agricultural research foundation. The foundation was being established with funds from the US government and facilities donated by the United Brands fruit company as it divested itself of its banana research in that country.[21] After visiting the research facility, I passed through Tegucigalpa and stopped at the US Agency for International Development (USAID) office to meet with government officials. Following the meeting, the rest of the team left for the rural area and I stayed behind in the city to work on the economics section of our report.

As I often do when traveling for work in a developing country, I spent the night in a small guesthouse—a sort of bed and breakfast. The owner had hired a young woman—let's call her Marta—to live in the house to cook and clean. I doubt she was paid much, but the job allowed her to feed and care for a young sister who lived there with her, and perhaps to send money to her parents and other siblings.

In the evening, Marta served me a delicious meal of rice, beans, and beef. Afterwards, I sat and worked on my project report in the living room, a ceiling fan barely stirring the sweltering night air. Despite the heat, Marta moved swiftly around the room, taking pains to sweep every nook and cranny. Barefoot, she wore a light sundress that barely hid her feminine features. After finishing her sweeping, she sat down across the room, fanning herself slowly. This was a little distracting and went on for about an hour, but we didn't talk as I needed to finish my work. At 10:00 PM, I said good night and told her I was headed to bed. At that point, Marta came across the room and quietly asked if I would like her to accompany me.

After an awkward pause, I said, "I am sorry, but no thank you." Marta said to me, "What is wrong with me? Is it my bare feet?" I assured her it was not—but was surprised by her pitiful appeal. She clearly needed money. I felt sorry for her, and feared that she may have thought I felt I was too good for her.

That night as I lay in bed, I couldn't help but reflect on a book entitled *Nectar in a Sieve* that I often assign my undergraduate class.[22] The book details the life of a poor peasant woman in India. At one point, her family suffers greatly when the rice crop fails. Her young grandson is starving, so her teenage daughter goes to town and sells herself as a prostitute for a small amount of food in an attempt to save him. It turns out to be too little, too late for the boy.

The next morning I packed my things to travel to another part of Honduras. Marta was in the kitchen, and she came out and walked me to the front gate to lock it behind me after I left, as she was expected to do. I thanked her for cooking and cleaning, and put $50 in her hand. She looked appreciative, but surprised, as she hadn't provided any "extra service." I felt good I had given her money as I walked away. I may have just encouraged questionable behavior by the gift, but perhaps she had a little brother or other relative who would eat better that night. In any event, if she was prostituting herself, she had to be pretty desperate.

Many of the world's poor in Latin America, Asia, and Africa scratch out a living on small farms, and most are better off than the poorest of the poor, such as the beggars in Bangladesh. Most of the poor may not have or produce much, but they work hard, and are surprisingly efficient at what they do. They have to be—their survival depends on it. It helps if their family has lived on the same land for generations and passed down agricultural knowledge over time. But risks are many, and the economic climate and external environmental conditions that affect farms are seldom static.

The one constant is change. Population growth, climate change, fluctuations in market demand for their products, increased labor costs, and new agricultural pests are just a few examples of changes that farmers face. New seeds, equipment, or farming practices may help them adjust, if they are available and appropriate, but improved policies, enhanced regulations, and educational opportunities are needed as well. Price policies that discriminate against agricultural producers, regulations that hinder adoption of improved technologies, and limited access to secondary schools are concerns. Many if not most of the poor farmers are women, so any changes in technologies or rules must also be made with an understanding of cultural conditions and financial, temporal, and other constraints that affect the adoption of new technologies by women.

I once met a woman—let's call her Afia—who lived on a small farm in Ghana with her husband and three young children. Her husband's father, two brothers, and the brothers' wives and children lived in the same compound. Household structures are complex in Ghana, with some being nuclear two-parent families, a few single-parent families, and many extended families that include a range of kinfolk, especially in rural areas.

Afia's house was made of mud with a thatched roof. A bundle of firewood lay against one of the outside walls. She rose at daybreak to care for her children, cook for her family over a smoky woodstove, grind maize, fetch water, wash clothes, gather firewood, and work in the field, among other tasks. She often toiled until after nightfall.

One day I passed near her village in the middle of May and spotted Afia bent over and barefoot, moving quickly across a field. Dressed in a long checked skirt and simple white blouse with her hair tied up with a scarf, she was dropping cowpea seeds in small holes she made with her machete. Her sisters-in-law were planting seeds beside her. Her young nephew was bent over hoeing another field nearby. Children of the three young women sat quietly under a tree with a six-year-old watching the younger ones, two of whom were babies. It was hot, but no one seemed to notice. The first rains had come, and it was important that the field be planted quickly to take advantage of the seasonal rains that normally would follow. If the rains arrived on schedule, the harvest would be good, and there would be cowpeas to share or sell. If not, the following months would be difficult, and the hungry season would be longer than normal. The most difficult days would come just before harvest the following year.

Afia has been fortunate. Rains have arrived on schedule in recent years, and none of her children have perished. Children in her village have died during abnormally dry years—simply too weak from hunger to recover when diarrhea hit. In Ghana and elsewhere in the Sahel, drought and diarrhea mean people die. Weather, poverty, and hunger intertwine. Those living close to the margin really suffer when the rains fail or some other calamity strikes. Fortunately, fewer young children die in Ghana today than 20 years ago, due to modest improvements in incomes and safety nets. But about 12 percent still succumb by the age of five, which is about average for a low-income country. The rate in the United States is about 1 percent.[23] After talking to Afia, I asked permission to take a picture of her and her family. She said yes, if I would send a copy back to them. I often take pictures and send them back to the subjects a few days later. The poor have few pictures, especially of their children, and are usually quite appreciative of the photos.

Farmers like Afia have limited resources, but are proud when they feed their families on only an acre or two. They wish they had fewer health risks and more opportunities for their kids. They mistrust politicians and are often deeply religious. They send their kids to school, where schools exist, and help their neighbors in need. Their children play happily in hard-packed yards around their houses. These kids have never heard of a PlayStation or Barbie, but have fun kicking a scruffy rag ball between goalposts made of sticks.

A fifth of the world's population is extremely poor, which is defined by the World Bank as making less than $1.25 per day, and 70 percent of them live in rural areas. Many are continually poor, especially the landless, the elderly, the orphaned, and the disabled, but others move in and out of poverty due to weather, violent conflict, pests, and other temporary misfortunes.

Most of us in the developed world are fortunate, not just because we have a decent standard of living, but because when we see human misery, "it passes from mind when it passes from sight."[24] The ability to forget— or grow accustomed to—misery is a blessing and a curse. It helps keep us sane, but also can lead to indifference. Many people care about their neighbors, near or far, but their actions, or inactions, may not reflect that care. That is why when an impoverished Julio and Imelda donate a pig to a pitifully poor neighbor, it gets to me. Julio and Imelda saw a need and they acted on it, with the little they had to give.

2

Fertility Fears

The world population has soared over the past 60 years, with much of the growth in poor countries. There are seven billion of us now, with at least two billion more projected before 2050. Some people worry about whether there will be enough food, water, and energy to go around. Fortunately, population growth *rates* have slowed in recent years. Should we be concerned about the number of babies being born, and what factors influence population growth rates? The highest population growth rates today are found in Africa, although population density is higher in Asia, especially in countries such as Bangladesh.

Bangladesh is a delta with fertile soil deposited through the ages by the convergence of two major rivers that pour out of the Himalayas into the Bay of Bengal. An appropriate moniker for the country might be "fertile flats," given its deep rich soils and its proliferating population, most of which lives on a flat river delta. One hundred and sixty million people, squeezed into a country a little bigger than Virginia, manage to survive, or not, by producing rice, vegetables, jute, and shirts—and by breaking bricks. Due to the dense population, the average farm size is less than an acre.

Housing in the capital city of Dhaka is difficult to describe. It is a mixture of semi-completed high rises and rows of decrepit huts. When I was a child, I grabbed whatever scraps of wood and tar paper I could find and built a "hut" in an old apple orchard on our farm. I camped out there with my brothers, but mostly used it as a fort during brutal games of apple dodge ball—a well-thrown green apple can really hurt. In Dhaka, it is as if hundreds of skinny kids got together and made huts—in a low-lying smelly field—and got all their relatives and friends to camp out there with them.

The brutality in Bangladesh doesn't flow from apple fights, but from floods. During the four-month monsoon season, it rains like rats and roaches—a more apt simile than cats and dogs as few of those survive the traffic while rats and roaches thrive. Dhaka is a city built with its back to a river. Flooding can be so severe that a sea of five-foot people in four-foot huts can wallow for weeks in water whose clarity rivals raw sewage.

Much of the rapid growth in world population has occurred in places like Bangladesh, where people seem packed in like shoppers at the mall on Black Friday. The good news is that population growth rates are on the decline in most developing countries, in part due to gradual income growth, which has reduced the need for as many kids. If your children are dying, you tend to have a lot of them just to make sure that a few survive to care for you if you are lucky enough to grow old. Years ago, when we asked women in Colombia how many children they had, they typically said something like 13, eight *vivos* (living) and five *angelitos* (little angels). Incomes have now grown and health care has improved in many countries. As a result, fewer children are dying than in the past, and the baby business has slowly stopped booming—just not everywhere yet.

In most countries, if a child lives to her fifth birthday, her odds of survival improve significantly—except in areas where malaria and HIV-AIDS run rampant, such as the tropical belt in Africa. Children are especially vulnerable at the time they are weaned. Their new food may not be nutritious enough or clean, and their immune systems have not had time to adjust to the new influx of bacteria. In rural areas, infant—and maternal—mortality can be especially high in part because most births occur at home, and there is limited access to medical care.

One day while we were living in Colombia, I rode my motorbike up the mountain to a rural store a couple of hours from town. The ride was difficult because recent rains had washed the road out in places, and I had to maneuver around piles of soil and trees that covered the road. When I encountered a large pile, I would walk my bike around it, being careful to avoid edging too close to the cliff that dropped down one side of the road. When I arrived at my destination, the scene was somber. A grieving father was in the one-room store with his brother. His 18-year-old wife lay deceased on the table, along with their child. When she had struggled in labor, the expectant father brought her to the store so they could catch the daily bus to town for help.

Unfortunately, because of the washed-out road, the bus never arrived. The brother of the almost-father and I spent a long quiet motorbike ride down the mountain to inform the authorities about the death. It was poignant for me as, at the time, Marj was expecting our son, John.

Fortunately Marj's childbirth experience was relatively uneventful. When the time approached for the birth, we traveled the long trip to Bogotá two weeks before the due date to be safe. Even then, a local custom made it interesting. When Marj went into labor, we entered the hospital at 7:00 AM and were ushered into a room. We were there alone until 5:00 PM, when a doctor entered and said it was time. Our son was born 15 minutes later. Marj and John were then brought back to the room, and we were alone once again. Because I knew little about what to do other than stare at our wondrous child, I decided to go ask why no nurse had looked in on us before or after the birth. The lady at the front desk looked puzzled. "Did you rent one?" she asked. It had not dawned on me that in Colombia the nurse might not come with the hospital. In rural areas of many poor countries, you are lucky to have a doctor.[25] Food and medicine may have to be purchased outside the hospital by family or friends and brought in to the patient.

Good basic health care can place downward pressure on birth and death rates. For example, even something as simple as oral rehydration treatments to reduce death from diarrhea eventually reduces birth rates. If there is no pension system and people expect a lot of their children to die young, they have many babies. Also, peasant farmers in remote rural areas who have limited opportunities for education, and need a lot of hand labor in the field, tend to have a lot of kids. If health and education systems improve, and the economy grows to provide employment opportunities with increased wages, people have fewer kids. As the availability of birth control increases, people have fewer kids. China is perhaps the only country with an explicit policy of only one child per couple, and that policy has significantly reduced its birth rate. Other countries are less coercive on birth control, but anywhere young girls go to school, delay marriage, and join the workforce, birth rates drop rapidly.

After the first two weeks at our Peace Corps site, Marj never had a single work day without an invitation to visit a home in the morning before her afternoon meeting with a women's group. She would discuss any topic that was on the mind of the woman of the home. She would give one-to-one instruction on macramé, crochet, knitting, pants making, or other topics. She might learn a new craft herself from the woman—which she would later pass on to her groups—and a lot about rural Colombian culture.

One day Marj visited the farm of Gloria and Jesus Giraldo. Their house was a little larger than some, as it had a kitchen and two bedrooms. Gloria and Jesus were both in their late twenties, had six children ranging in age from six months to eight years, and farmed about two hectares. They grew coffee, plantain, sugarcane, cassava, maize, and beans, and were relatively well-off compared to their neighbors. Two of their children were in school that day, but four of them were still too young, and three of them were very curious to see the fair-skinned blond lady sitting on their porch. Gloria told the kids to go play. As she fed her infant, she asked Marj if she could discuss something personal with her.

Marj had become accustomed to such conversations. The longer she worked with women's groups, the more individuals confided in her. It helped that Marj is such a great listener, and people valued her advice. This particular day, Jesus was out in the field and Gloria opened up to Marj. She said she was tired. She loved her husband and children, and liked caring for their needs. She enjoyed cooking, and didn't mind tending to the chickens, the pig, and their home garden. Washing clothes by hand was a chore because they were muddied so fast, but she was used to it. But she was concerned about her ability to handle more kids. She was worn out, and she worried that they might not be able to afford the school fees if they had more children. She felt school was becoming increasingly important, as there was little land left to farm. Her question for Marj was a simple one. How did Marj manage to be married for more than a year without having any children? Gloria had heard of birth control, but wanted to know some specifics. Marj responded to her as a friend, although somewhat carefully, as we were not there to teach family planning in a country where the prevailing religious teaching opposed it. In the early 1970s, family planning was not as actively supported by Colombian government policy as it is now.

It turned out that many Glorias started to ask that same question about that time. The annual population growth rate in the early 1970s in Colombia was almost 3 percent. But forces were unleashed that cut that growth rate in half and the birth rate by almost two-thirds over the next 40 years. Today, the population growth rate in Colombia and in much of

Latin America is only slightly higher than it is in the United States. What happened to cause such a substantial change? And what does it mean for countries and the world?

Several forces were at work: income growth; increased educational opportunities for girls; urbanization, which reduced the need to have children for labor; the spread of family planning; and improvements in health care that reduced child mortality and the need for large families for security during old age. What does the decelerated population growth mean for the future? Countries like Colombia will find it easier to employ their people and continue to raise per capita incomes. In addition, the flow of immigrants entering the United States from countries in Latin America should slow because the primary reason for the flow has been the search for jobs.

Forty years ago, there were dire predictions that an exploding world population would cause mass starvation.[26] For the world as a whole, those predictions were wildly wrong, in part because food production has roughly kept pace with population growth and in part because population growth rates are currently only about half of those in the late 1960s. Global population has nevertheless doubled since that time due to the large number of people in their child-bearing years and longer life spans in general. Unfortunately, some countries such as Guatemala, Pakistan, Uganda, Tanzania, and Burkina Faso have been slow to experience lower population growth rates. These countries have lagged in improving the food and income situations for their populations. As those situations change in the future, population growth rates may well slow down, especially if public policies encourage female education and family planning.

Ironically, while overall population growth is still rising rapidly, the sharp reduction in its rate of growth in some developing countries (such as China) also raises concerns about how they will eventually support their elderly populations. A few years after a country reduces its fertility rate, it receives an economic boost that lasts for several years because the large working age population has fewer dependents to feed and otherwise support. More money is available for savings and investment. Some of the economic growth in East Asia over the past 30 years was fueled by such a "demographic dividend." As that large cohort of working people eventually ages and is followed by a smaller working-age population, those countries must hope that their economic growth has been sufficient to provide an income base for the elderly. Otherwise, poverty rates may rise again. They are in a race to be rich before they are old.

3

Hunger Hurts

For hunger is a curious thing: at first it is with you all the time, waking and sleeping and in your dreams, and your belly cries out insistently, and there is a gnawing and a pain as if your very vitals were being devoured, and you must stop it at any cost, and you buy a moment's respite even while you know and fear the sequel. Then the pain is no longer sharp but dull, and this too is with you always, so that you think of food many times a day and each time a terrible sickness assails you, and because you know this you try to avoid the thought, but you cannot, it is with you. Then that too is gone, all pain, all desire, only a great emptiness is left, like the sky, like a well in drought, and it is now that the strength drains from your limbs, and you try to rise and find that you cannot, or to swallow water and your throat is powerless, and both the swallow and the effort of retaining the liquid taxes you to the uttermost.[27]

In times of famine, pictures in the media of people suffering from extreme hunger can be gut-wrenching. Famines result from insufficient access to food due to crop and political failures in defined areas and time periods. Famines are devastating, but produce only a small fraction of hunger-related deaths. Chronic undernourishment is more pervasive, with about a billion people suffering daily from the effects of living on less than 1,800 calories per day. Someone who is starving in a famine might be happy to go back to just being hungry, but for most people in low-income countries, chronic hunger is their greatest threat.

Fundamentally a poverty problem, hunger is particularly cruel to young children. Roughly a third of all children in the world are too light for their height—classic signs of malnourishment. According to the World Health Organization, one child dies every five seconds as a result of hunger. That such hunger occurs anywhere reflects poorly on humanity everywhere.

I often jog at daybreak. It opens a window to a world that can be closed at other times of the day: people scurrying to work, a child sleeping in a doorway, a man improvising a bathroom. It is 6:00 AM in Tegucigalpa, Honduras, and I jog through hilly streets, searching for an area with fewer open manholes. Rounding a corner I see two little boys, one devouring breakfast he had just scrounged from a garbage can, the other still sifting through the trash. They look dirty and hungry. I jog by with a tinge of guilt.

It is 6:00 AM in Manila, the Philippines. I am jogging along, admiring the view of Manila Bay. Four thin teenagers surround me, admiring my watch and Nikes. Adrenaline kicks in and the shoes impart surprising speed as I escape the skinny teens. No wonder they preferred the shoes to their flip-flops, but I feel no guilt in keeping them.

It is 6:00 AM in Dhaka, Bangladesh. I jog by a woman and her two young children in front of a blue plastic bag lean-to. They look hungry. One little child, dirty and naked, is squatting in the street, using it as a toilet. I keep running, and it hurts me. Hunger tears at heartstrings as well as stomachs.

It is 6:00 AM in Blacksburg, Virginia. I jog along a quiet street, passing a runner, a cyclist, and three people walking their well-groomed dogs. The early fall air is crisp and clean, the leaves just starting to show hints of orange and maroon. My thoughts escape to Hokie football.

The generosity of those with little to eat often amazes me. In Colombia, Ecuador, Senegal, the Philippines, and even Bangladesh, if they know you are coming, and often even if not, they feed you. Not always the best food, mind you, but you do not starve, even figuratively. Walk onto a farm unannounced in Senegal, and they give you peanuts. If you stay around very long, they give you a spoon, and soon you are eating rice and vegetables out of a common bowl with the family—or at least with the males in the family. In the *campo* of Colombia, you had better like chicken if they expect you for lunch. They will kill it and cook it while you wait. Sharing is strongly embedded in the culture. In the Peace Corps, I was fed so many *arepas* that I had to sneak some out in my hat so as not to insult the generosity.

One of the poorest rural neighborhoods where we worked in Colombia was called El Congal. Marj organized a women's group there and visited individual homes. One family she visited had 10 children, one of whom did not attend school because her bones were deformed and she struggled to walk. Marj was sure the child had rickets, and she suggested ways that the mother might use the powdered milk given to her in a food package at the local health center.

Marj visited the family several times and helped teach the girl to read. The first time she went, the family insisted on feeding Marj even though they didn't have enough food for themselves. They gave her rice and a piece of fish skin. She felt bad about eating it—not because it smelled fishy, but because it meant the youngest kids only got a small portion of rice. Marj ate what she was served, however, because the mother was so pleased to have her eat with them. She said the previous female extension agent who had visited would not even come into their house, much less sit down and eat with them.

Despite trying hard to avoid being in hungry homes at meal time, Marj and I can't count the number of times we were offered food by families whose children appeared ragged. We always felt guilty accepting it, but usually did, not wanting to deny them the pride they felt when giving to a guest. It is a humbling experience to accept food from the poor, one that can take more courage than giving it.

People are generous, but I have to admit that eating food given to me in remote areas sometimes takes guts. The common bowl in Senegal usually has flies buzzing over it. Participating diners wave their spoon-less hand in a sort of rhythmic fly-shooing motion throughout the meal.

In southern India, the food is so spicy that my host once thought I was having an unhappy meal because I cried so hard while I ate her delicious food. If I just hadn't mistaken those darn little green peppers for string beans. . . .

And, a poor family in Colombia once gave me a well-cooked piece of meat that smelled borderline rotten. I thought I'd be clever and sneak it to the dog under the table. But he took one sniff and refused it. I then had to figure out how to slip the dog-sniffed meat into my clothing so as not to insult my host. My hat proved handy.

Hunger and culture can intertwine in many ways. I was having dinner with the director of the Philippines Rice Research Institute (PhilRice), an outwardly gruff, but kind-hearted, old man named Santi. As we ate, he recounted a surprising story involving his dog and a field-worker at the PhilRice research station. Santi had a dog that constantly stuck by his side. He was only a mutt, but Santi loved that pooch and some days he even took him to work. One day Santi brought him to the office, but during the day was called away to a meeting. He let the dog loose to wander on the station grounds while he was gone.

When he returned, the dog had disappeared. He asked everyone if they had seen his dog, but the animal was nowhere to be found. Someone thought he saw a worker leave with the dog, but he was not sure who it was, and the workers had already gone home for the day. Santi was very upset and decided he would confront the workers in the morning.

After dinner that evening, there was a knock on the door. A sheepish-looking man was standing there with a little puppy. The man explained to Santi that he was a field-worker at PhilRice, and he had been employed there only a short time. He didn't have much money and the job was very important to his family. Even with the job, his family was barely making ends meet. He had seen a stray dog at the station that day and had taken it home. His family didn't have much food and so they had eaten the dog for dinner. He had no idea it was owned by the director. Someone had told him that evening that the dog he had taken was Santi's. He was extremely sorry. He had brought the puppy to Santi to make up for eating his dog and he hoped that he could keep his job.

Santi was devastated to learn what had befallen his pet. He was very upset and he reprimanded the worker for stealing a dog whose ownership was uncertain. He found the idea of eating a dog revolting, although it is not uncommon in the Philippines. He told the worker he would decide his fate in the morning.

That night Santi could hardly sleep. The man had stolen something, and he had committed a cruel act. And of course Santi missed his dog. But the man was clearly desperate. The next day Santi let him keep his job, but made it clear that if he or anyone else was ever caught stealing again, he or she would be fired. He also decided that he would treat his workers more often to a highly nutritious local delicacy called *balut* (fertilized duck eggs)—and he kept the puppy.

Severe hunger is seldom constant. If it were, those affected would die relatively quickly and constantly over time. During famine, people indeed die quickly, but famines are of short duration. In periods without famine, which is most of the time even in the poorest countries, people reach a food-energy balance for months or years that helps them to survive from crisis to crisis. Many rural areas, especially in Africa, have hungry seasons, the weeks or months before harvest when money and food stocks are low but the grain is not yet ripe. The quantity and diversity of food consumption declines, and children display distended bellies and other signs of malnutrition. Hungry seasons turn deadly if the previous harvest was poor, or worse yet, if the two previous harvests failed. In many developing countries, hunger is like a predator that lurks just below the surface and reappears periodically to eliminate the weak.

When crops are destroyed by drought, flood, or pest, less eating is done that year. People have many strategies to minimize the risk of hunger, such as diversifying crops and livestock, selling assets, eating "bush meat," storing food across years, cutting back on school fees—and schooling—for children, and seeking assistance from relatives and friends, but these strategies also fail when subjected to extreme or serial calamities. A solution to the problem is to find means to help people broaden their strategies, including local off-farm employment and adoption of risk-reducing agricultural technologies.

Not all hunger results from a food-energy imbalance. Why might people be out of sorts even when they have a reasonable amount of energy and protein? The simple answer is: they lack vitamins (especially A and E) and minerals (especially iron and zinc). In impoverished rural areas, it is common to find people with blindness, stunted growth, pneumonia, impaired activity levels due to diarrhea, and other problems related to micronutrient deficiencies. Many children under five as well as pregnant and lactating women suffer and die from these deficiencies.

Let's face it; hunger is horrible. If we had to choose between dying from obesity and dying from hunger, most of us would pick obesity. Fattening takes longer, and it is definitely more fun gorging on pizzas, burgers, and fries than it is fretting over a lack of food. Just the thought of starvation makes us squirm—and not because it reminds us of dieting. We know it is a lot more serious, and feel helpless to stop it.

What can be done about extreme hunger and malnutrition? One obvious solution is economic development to improve incomes. But some countries, even at the same low level of income, perform better than others in improving nutrition and health. They have basic safety nets in place involving emergency food distribution centers and plans, preventive health care, and conditional cash transfer (CCT) systems that provide money to households if their children attend school or visit doctors (or both) on a regular basis. Mexico was one of the first to implement such a program in the mid-1990s.[28] Brazil followed with what is now the world's largest CCT program as part of a broader effort to reduce malnutrition. The effect on nutrition and health indicators in Mexico and Brazil has been significant.[29] Other countries, especially in Latin America, but also in Asia and Africa, have taken notice. Some, such as Colombia, the Philippines, and Bangladesh, have initiated CCT programs and are beginning to see the effects.

Increased food production is important but not enough to significantly reduce extreme hunger. Bangladesh has tripled rice production over the past three decades and its infant mortality rate and other measures of malnutrition have declined, but a third of Bangladeshi children remain undernourished. Malnutrition has been frustratingly persistent in India as well despite recent economic growth. Basic safety nets for the poorest of the poor are essential, as are the provision of micronutrients (vitamins and minerals) and macronutrients (calories, protein, and fat).

4

Feeling Sick

My graduate student, Pricilla, was conducting a survey of farm households in Nigeria when she became feverish, then chilled, began vomiting, and suffered a severe headache. She went to see a doctor, who told her she had malaria. She was surprised as she had been taking the daily preventive pills, but it hit her anyway—the same lament I suspect doctors often hear from pregnant women. Prophylactic pills do not always prevent malaria. Pricilla was fortunate because the doctors in Nigeria are quite familiar with malaria; in fact, about 25 percent of malaria cases in the world occur in that country. They gave her a treatment that quickly helped and within a few days she was able to resume her research. She had a headache that lingered for a couple of months, but all in all she was fortunate—and is pretty tough, too. She also had the advantage of being well-nourished.

Health and hunger problems are tightly connected, and not just because they both occur among the poor. Infections and parasites lead to malnutrition, while malnutrition impairs immune systems, increasing the risk of infection and severe illness. Illness reduces appetites, causes malabsorbed food, or results in nutrient wastage due to fever and other metabolic processes. Some illnesses lead to diarrhea, which in turn can lead to dehydration and death. Malnutrition and poor health also reduce the ability to earn incomes, so when poor people become ill, they often become trapped in a downward spiral from which they struggle to escape.

Malaria is one of the most serious problems, especially across the warm belt in Africa and South Asia. Malaria parasites are transmitted from person to person by mosquitoes that mostly bite at night. If the disease goes to the brain, it can be deadly, especially for young children. According to the World Health Organization, malaria afflicts as many as half a billion people each year, killing roughly a million. Malaria can also invade the digestive tract and block nutrients from traveling to the brain. Combined with diarrhea, it can seriously rob the brains of young children of critical energy and protein, impeding mental and intellectual development. It also causes a significant loss in labor productivity, which contributes to poverty.

I have been fortunate to remain free of malaria symptoms, despite extensive time spent in areas where the disease is endemic. I take preventive medicine, but the major reason I have not contracted a significant case of malaria is likely the portable net I carry with me and drape over my bed and tuck under the mattress at night. Malaria-carrying mosquitoes typically bite after dark, and I have fallen asleep many nights to the stereophonic sound of mosquito wings just outside the net. Relatively inexpensive bed nets have been proven to be an effective way to reduce malaria incidence. No anti-malaria vaccine is available yet, and it is difficult to reach everyone with preventative medicine for a whole lifetime. Malaria treatments recommended by the World Health Organization include therapies that use multiple drugs to reduce the problem of malaria strains developing resistance to the drugs. Until a malaria vaccine is developed, distribution of free or low-cost bed nets appears to be the most effective way to reduce the disease.

Malaria is just one of many health maladies that affect life spans in the developing world. Until the middle of the nineteenth century, average life expectancy globally was only 30 to 40 years. But in the century and a half after that, life expectancy doubled in today's high-income countries as medicine and improved sanitation controlled many illnesses. Unfortunately, life expectancy continues to lag dramatically in many poor nations. If you are born in Africa today, your chances of living to age 50 are about 50 percent. In contrast, in the United States you have a 50 percent chance of living to age 78. The high death rates among infants and children pull the African average down.

Through concerted efforts by governments, foundations, and private industry, significant progress has been made in recent years in reducing health problems, even in some of the poorest nations, but many problems remain, big and small. For the poorest of the poor, death lurks behind every drink of dirty water, untreated cut, or mosquito bite. The most vulnerable are young children, especially right after weaning when their immune systems are still adjusting. It is difficult to visit a developing country without being struck by the pervasive nature of health concerns, acute and chronic.

I arrived in Bogor, Indonesia, near the end of the rainy season and in the midst of a dengue fever epidemic. Had I known about the dengue, I might have found an excuse to stay home. No illness is fun, but dengue fever, also known as hemorrhagic or breakbone fever, can be really nasty: severe headache, debilitating muscle and joint pain, vomiting, diarrhea— and even worse if you count dying. Only a small percentage of its victims died during that particular outbreak, and they were mostly the weak— the very young, the very old, and the otherwise infirm. But people were dropping like—well, like people.

Dengue fever shares with malaria its transmission by mosquitoes and widening geographic spread in recent years. Scientists are working on a vaccine and on means to control the mosquito that spreads the disease, but solutions appear to be a few years away.

I have never had dengue, but on one trip to Bangladesh, I did become ill and passed out on a plane as it was about to depart Dhaka. When I came to, the passenger next to me was taking my pulse and the flight attendants were hovering. Our airbus was poised at the end of the runway blocking traffic while my situation was being resolved. The pilot wanted to know if he should take off for Singapore or go back to the terminal so I could be admitted to a local hospital. Reflecting for a second on the relative sanitation levels in the two countries, I thought it best to ignore my light-headedness, profuse sweating, and barfing, and told the pilot to gun it. Good choice. By chance, the passenger sitting next to me was a nurse who tended to my medical needs during the flight to Singapore. Sometimes an angel appears at just the right time.

The cleanliness of medical facilities and safety of blood supplies in numerous countries (such as Bangladesh, Tanzania, and Zimbabwe) give one pause, as it did to me that night on the runway in Dhaka. It is hard enough to avoid infections in a US hospital. It is that much tougher in places where you cannot drink the water and where illnesses such as hepatitis and AIDS are rampant.

The relatively young director of the National Agricultural Research System in Tanzania rose from his chair behind a large desk in his spacious office in a dingy government building on the outskirts of Dar es Salaam. I noticed how tall and slim he was as he crossed the room with an unsteady gait to greet my Tanzanian colleagues and me. He asked us to sit with him at a large table. We were there to discuss the analysis of agricultural research priorities that we were completing at the request of the Tanzanian government.

I have visited with research directors in other countries during similar analyses. The discussions have always been lively, given the sensitive nature of the topic. But on this day, the director was passive, almost like a student in my 9:00 AM class after a Thursday night football game. He would ask a question, but he seemed in a daze, his mind wandering. Later, I asked what the story was with the director. They told me that he was an excellent director, but that he was ill. His hometown was in the northern part of Tanzania near Lake Victoria, an AIDS hot spot at that time (1989). The director had good days and bad days, and they were numbered.

Leaving this sad situation, the next day we continued our work, which required visits to research stations scattered around the country. We traveled by SUV on roads with more potholes than pavement, through incredible wildlife preserves, and across wide open spaces. Traveling south to Mbeya, we spent the night at a local lodge along the way. The next morning as we drove back to the main road, we encountered an elephant blocking our narrow path. He flapped his ears—not a good sign—and charged. With tall grass on both sides of the road, we had no choice but to challenge the speed record for backing up in an SUV until we found a spot to pull off the road. The elephant ran on by, fortunately.

I don't know why, but I seem to attract animals. In the Peace Corps, I was bitten by dogs four different times. I never saw two of them coming, and the other two disproved the adage that barking dogs don't bite. I was a little concerned about rabies until the owners told me not to worry; their dogs bite everyone. One dog crawled under several seats on a bus to sneak up behind me and bite my leg. And my propensity for animal clashes is not limited to developing countries. One day I was jogging down the road in Blacksburg, minding my own business, and got hit by a deer. I learned that deer are very solid animals.

Back in Tanzania, the day after the elephant incident, we traveled north to Arusha. When we stopped for supplies, I spotted a group of Maasai walking near the road with their cattle. The Maasai are tall, colorful folk, so I had the bright idea to snap their picture. Not so bright. Objecting to having their picture taken, they chased me back to the car, waving their long staffs like spears. Once again my Nikes came through. And at least I did not get hit by one of their cows.

The next day, we visited wheat and coffee research stations and then began a long journey back south. Rounding a bend about nightfall, we came upon an overturned tanker truck with several people lying in the road. None were dead, but some were injured and bleeding. We managed to load all six into the SUV and took off for the nearest town, about two hours away.

I found myself torn. They were drenching our vehicle with blood, and I had seen way too much evidence of HIV on this trip. To make matters worse, two of the victims kept beating on the semiconscious truck driver, blaming him for causing the accident. They had been hitchhiking and he had allowed them to ride on top of the tanker. But the driver had been drinking and had sideswiped another vehicle on a curve, resulting in the accident. Afterward the hitchhikers were out for more blood. And if the woman with the baby would have only stopped wailing, her child might have stopped competing for shrillest shrieker on Tanzania's Got Talent.

After dropping our passengers at the hospital and filing a police report, we stopped to eat. I was careful to wash any blood off my hands, but was not too hungry anyway as the dirty water had caught up with my stomach that day. As we sat there nibbling our food, one of my Tanzanian friends said to me: "You seem more like a Peace Corps volunteer than a professor." It was intended and taken as a compliment—although I suppose it doesn't say much about professors.

In areas where transportation limits access to a health care facility, even a relatively minor injury can prove life threatening. One day in Colombia, Marj and I motorbiked out of Pensilvania to meet with a group of farmers in a remote area. About 10 miles out of town we met a man in the road stumbling toward us with an axe. He was totally drenched in blood and looked like he was auditioning for a Halloween horror flick. It turned out our Freddy Krueger[30] wannabe had been carrying a load of wood down the mountain when he fell and cut the back of his head open with the axe. Blood was still streaming from his wound. That he was still coherent amazed us. We wrapped his head with cloth and helped him climb onto the back of the motorcycle for the trip to the hospital in Pensilvania. He was pretty groggy when we arrived, but he had somehow managed to hang on behind me as I sped around the curves in my motorcycle ambulance.

Freddy was fortunate that day, as he might not have encountered another vehicle on that dirt road for hours. Injuries present serious problems in remote areas with few real ambulances, or even vehicles for that matter. At least today, the increased prevalence of cell phones is beginning to help.

A Colombian horse trader we knew was not as fortunate after an encounter with a venomous snake called a mapaná, or fer-de-lance (*Bothrops atrox*). I first learned about this scaly serpent from a personal experience with one on a path just below the coffee zone in Colombia. I was descending a mountain trail with Luis Carlos when suddenly a long, thick snake with a pointed head crossed about a foot in front of me. In Colombia, it was not uncommon to see small snakes such as the coral, with its distinctive red and black bands, pass in front of us as we walked along a mountain path. They always slithered away quickly, preferring to mind their own business. Perhaps it was its dark color, the X-shaped markings along its back that seemed to shout poison, or the fact that it was as thick as a wrist, but this mapaná seemed considerably more menacing.

The snake did not strike, but after it passed, Luis Carlos gave me the scoop on its usefulness and personality and what to do if bitten. This pit viper does help farmers by consuming the ever-present rodents in grain fields. While its bite is deadly if untreated, the venom usually takes several hours or even days to kill, so the victim has time to seek medical help. We were wise to give the snake space because it has an irritable disposition, is easily agitated, and strikes quickly with little provocation. The bite is big and painful.

About a month later, I was sitting in a café in Pensilvania negotiating the purchase of a horse. I had seen the beautiful painted mare the week before and was immediately struck with new horse fever. During our negotiations the horse trader pulled more tricks than a used car salesman, including switching the horse I had test ridden for another similarly marked but smaller beast. I closed the deal before he could try to sell me rust proofing. While this horse trader did not endear himself to me, I was sorry when he was brought back to town in the Colombian equivalent of a body bag about a week later. He had surprised a mapaná in his pasture and it struck him in the leg. Alone, he made a fateful decision to cut the wound with his machete and tried to suck the venom out of his leg. He bled to death before he could make it back to town. If he had just traveled back to town without cutting his leg, he probably would have survived.

One day, I helped a Colombian farmer named Joaquin establish a vegetable garden just below his modest two-room house. As we worked, he told stories that made his kids and me roll with laughter. He didn't have much, but he sure seemed to enjoy life.

A couple of months later, I returned to see how his garden was growing and was met at the door by his wife, dressed in black and very somber. She proceeded to tell me that Joaquin had not felt well for some time, and a few weeks earlier had suddenly become very ill. She took him to the doctor in town, but he was so sick they sent him by bus to Manizales, the closest city. He died in the hospital there. The doctor told her that he had cancer.

I gave her my deepest condolences, or at least I tried to after racking my brain for the right words in Spanish and only coming up with: "*Me da mucho pésimo,*" which translates something like "I feel very bad." I was at a loss about what more to say to her and asked if there was anything I could do to help. She suggested I weed his garden, as he had been so pleased with it. I think she was trying to cheer me up. It was a sad weeding.

Marj is fair skinned. She spent her summers in college as a lifeguard and swimming instructor at a beach on the St. Lawrence River. One day several years ago, she noticed that a mole on her arm had turned black and crusty. She went to a dermatologist who removed it in a hospital the next day. Tests revealed that my river-watch babe was lucky. The cancer had not spread to her lymph nodes. Melanoma is dangerous. If it is caught in time, the cure rate is close to 100 percent. If it has spread even a little, the chances of survival are slim.

If you live in a poor country and contract melanoma or any other cancer, your odds are bleak. Except for the wealthy, people often are less aware of cancer symptoms than in the United States, find it harder to locate a cancer specialist, and struggle more to afford treatment. Even in the United States, people living in poorer rural areas often lack access to cancer treatment. In southwest Virginia, the breast cancer rate among women is lower than in the rest of the state, but death rates from breast cancer are much higher. Many women in Appalachia do not realize they have cancer until it is too late for treatment.

Of course not all health problems are acute. One of Marj's women's groups in Colombia had a chronic problem. Sugarcane was grown locally and it was cheap. As a result, folks were accustomed to a little *agua de panela* (boiled brown sugar water) pick-me-up several times a day. Besides hyper kids, the predictable effect was that most adults could pass for NHL hockey players—missing teeth, loose teeth, and brown teeth. The missing teeth may not have helped their sex appeal, but the real problem was the loose and brown ones. They ached badly. In desperate need of a dentist, the women pleaded with Marj to cajole the one dentist in the region to pay a visit to their community. He agreed to come for a day if Marj would play dental assistant. It was a tough day for her.

The "*dentista*" arrived with a small bag. After breakfast, a straight chair was set on the front porch, and Marj was given 30 seconds of *dentista* assistant training. Her task was simple. Ask the patient to sit in the chair. Stand behind the chair. Hold the patient down by exerting steady pressure on the top of the shoulders. About 60 patients lined up behind the chair. The dentist opened his bag and pulled out a pair of pliers. No Novocain, no free toothbrush, just a good old-fashioned tooth pull. With little talking and no complaining, each patient sat down and stoically endured five to 10 minutes in the chair. The dentist sweated a few stubborn molars, but most teeth came out with surprisingly little effort. One tooth was removed by a finger flick to the tooth—no sense wasting time on pliers for that one. The tougher ones required a little more shoulder pressure on Marj's part; she was closer to tears than were the patients, even the kids. The dental clinic lasted all day. Each person was very appreciative of the dentist and of Marj. Ever since that day, a visit to my dentist has seemed like a piece of cake.

Dental health is one of many chronic health issues that affect productivity among the poor. Poor eyesight is another. I observe only a tiny percentage of children and adults wearing glasses in developing countries, especially in rural areas. The number of kids I see squinting tells me that the absence of glasses is not due to a good contact lens salesman. Their eyesight is really poor, sometimes from the normal nearsightedness experienced by about one-third of most populations, and often from vitamin A deficiency. According to the World Health Organization, up to half a million children in developing countries lose their sight each year for lack of vitamin A.

Chronic illnesses present other challenges, such as limited accessibility for the handicapped, in rural areas of developing countries. Marj would love to continue to accompany me on trips to Bangladesh, India, Ecuador, China, and other places where textiles and apparel constitute a major industry. But her MS is now just too constraining for such travel. She tires easily, becomes stiff, and struggles with her balance.[31] Handicap access is quite limited in most developing countries.

Here we often take for granted the benefits of a relatively accessible environment for living, transport, and work. I mentioned in chapter 1 that the youngest son of Julio and Imelda Ospina had polio. He could barely walk, even with assistance from his parents or siblings, a major disadvantage given the long steep hike through a pasture from the road to their house. Marj has had MS her whole professional life, but with perseverance, assistive devices and ramps, and faithful friends, she has been able to adjust to each new challenge of the disease, be a productive professor, and raise a great kid. Once when I was overseas, Blacksburg was hit by a major snowstorm. Without asking, a colleague, two of our neighbors, and the driver of the town (handicap) access bus shoveled our driveway, so Marj could get out on the bus to teach her class.[32]

In a rural area in a low-income country, people may be helpful, but their local conditions make it nearly impossible to meet a challenge like MS as effectively as my wife has in Blacksburg. For someone with such a chronic illness, the result can be great dependence on others and a short life span. I recently visited a farm in the mountains of Ecuador where a 30-year-old woman in a wheelchair was basically trapped in a small house with only her elderly parents to care for her. The parents went on at length about how difficult it was to get her to and from the bus so she could visit the doctor in town.

Inability to afford medical treatment, especially for people with chronic illnesses, is another pervasive problem for the poor. I visited a home in Colombia several years ago and found a woman who had been in bed for 10 days following a heart attack. She struggled to move, but no one had called a doctor because they could not pay. Even if she could have afforded treatment, the journey over a rough path for two hours to get to the road to catch a bus to town would have been risky.

The health care situation is not uniformly better in developed countries. Basic health care in some middle-income developing countries is now almost as available as and less expensive than in the United States. When one of the undergraduates I took to a rural town in Ecuador became ill and dehydrated, I called a private doctor. He charged only $10 for a house call in the middle of the night. Unfortunately, in parts of Africa and Asia, medical care is much harder to find and $10 may be a week's wages. In the United States there is one physician for every 39 people. In several African countries, there is one physician for every 3,000 or more people.[33]

If I had taken the undergraduate to a public hospital in Ecuador, it would have cost even less, but the wait for someone with a problem that was not immediately life threatening might have been substantial. To a significant extent, private doctors and health facilities are there for emergencies and the well-off. Public facilities are for the rest. The good news is that more and more countries are developing multiple-tier health care systems so that there is at least primary health care for most people, even if many of them have to wait in line for a long time.

An observation, also borne out by statistics, is that nations with a strong emphasis on primary health care have the best health outcomes. A strong primary health care system requires a sufficient number of trained health care workers with incentives to do their jobs, health care leaders who are accountable, community participation in local health centers with a range of services, coordination when people need referral to specialized care, and low financial barriers for the poorest people to access assistance.[34] Tasks in the system should be assigned to the least-expensive health care worker who can complete the task reliably. When people are forced to go to hospitals for maladies that could be treated at a local health care center, the system tends to break down due to cost and insufficient capacity. In areas without a local health care center, people often go to the local healer or herbalist, with predictable outcomes.

5

Fragile Fields

I once observed a farmer in the Central Luzon region of the Philippines spraying pesticides on an eggplant field. It was hot and he was wearing a light shirt, shorts, and flip-flops. The sprayer was mounted on his back, and as he pumped a handle at his side, the spray drenched not only the leaves in front of him, but his hands and legs as he walked down the rows. When he reached the far end of the field he stopped to relieve himself of the liquids he drank for lunch, undoubtedly covering his sensitive body parts with pesticides in the process.

When he came back to my end of the field, he stopped and mixed chemicals from several different bags into his sprayer before adding water—with roughly the skill of a six-year-old helping his mother bake cookies. Then, with his hands covered in an assortment of pesticide powders, he stopped for a cigarette, leaving a white chemical mustache on his upper lip after a few puffs. He could have starred in a "Got Pesticide" commercial.

I figured he was spraying for insect pests, particularly a fruit borer endemic in the region, but I asked why he had chosen the particular chemicals he was using. His response: "After spraying the eggplants with this mixture, I feel woozy, so it must really be working on the insects."[35] I resisted the temptation to tell him that the pesticides might be clouding his judgment, and decided to speak to someone in the Philippine government about the need for pesticide education—and regulation.

It is easy to forget that not too long ago—before Rachel Carson's *Silent Spring*—we were pretty lax in the United States about pesticide safety. For example, when I was a child, we often hired a person with a sprayer mounted on an open Jeep to come spray our corn fields for weeds. I would beg my father to let me ride along. What I really liked was the way the spray drenched me when we swung around a corner in the field. The chemical was kind of sweet tasting. Maybe that explains my bad jokes—a little too much 2-4-D. Fifty years ago, farmers in the United States were less aware of the dangers of farm chemicals.

I trekked around Nepal for two months on an assignment for the Food and Agriculture Organization of the United Nations, visiting agricultural research stations as part of a review of the agricultural research system in the country. Nepal is a spectacular country to visit with its diversity of topography and climate—rows of snow-capped peaks overlooking steeply terraced slopes and vegetation that varies from arctic to tropical. All of this diversity can be observed from one spot on a mountainside. One day our small group traveled from the city of Pokhara up those picturesque slopes until we ran out of road. We spent the rest of the day walking up a path on a gradual incline until we reached a research station called Lumle, established by the British decades earlier.

As we huffed and puffed our way along, we were frequently passed by Nepalese women almost sprinting up the hill. They carried everything from baskets of vegetables to 20-foot lengths of plastic tubing on their heads. After a while, I discovered one reason they passed us so quickly. Although the path was inlaid with stone, dirt filled the spaces between them. The first person walking along stirs up leeches hidden in the soil, which then latch on to the ankles of the next person who follows up the path. The leeches inject a numbing agent as they suck blood so you don't feel them. I didn't even notice until someone told me that my ankles were all bloody. The crafty women knew to never walk close behind anyone and to always pass quickly.

As we continued our trek, our Nepali host talked about the nuances of the farming we encountered. At one point, he gestured toward the mountainside across the valley from the one where we stood. There a sizable landslide had occurred. It had carved out all the soil from the pasture high up the mountain to the carefully terraced hillside down below. The soil had given way and all that remained was bedrock.

Our Nepali friend pointed to a school on the side of the mountain right beside the landslide. He told us that the side of the mountain had given way about 8:15 AM on a school day. Just 15 minutes earlier the kids had been walking to school along the path wiped away by the landslide. A child who was late for school that day had perished in the slide.

Why did the slide occur? Population growth and poverty had forced people to cut down the forest for firewood and to farm the cleared land high up the steep slope. Crops that replaced the trees were incapable of holding the saturated soil once the heavy rains began. Soil erosion and landslides are common problems in many poor countries where farmers worry more about feeding their families in the present than about erosion or landslides that may or may not occur in the future. The problem is compounded downstream after eroded soil is deposited as silt in the rivers. The silting causes flooding not only in the low region of Nepal, but also in India and Bangladesh.

The government of Nepal has worked to reduce incentives to deforest the land by encouraging community-based natural resource management and by developing alternative sources of energy, as much of the harvested wood is used for fuel. For example, simple methane generators have been developed to turn animal manure into gas for cooking while leaving compost for fertilizer. Progress is gradual, however, and much remains to be done.

I traveled with a soil scientist to visit Yurimaguas, a river port town in the northeast Amazon region of Peru. We poled along in a dugout canoe downriver until we came to a small clearing. Leaving the canoe behind, we walked on a path through the jungle until we arrived in a newly opened area planted with rice. Most indigenous farmers in the area grow a mixture of crops together including cassava (a root crop), plantain (cooking bananas), maize, sugarcane, and others. It is hard to distinguish their farms from the forest. But cutting down a large swathe of the rainforest to open an area to grow rice is relatively new, and its long-term environmental consequences are uncertain. The slash and burn method used to clear the land releases carbon into the air, and the ability of the soils to sustain rice monoculture may be limited. The population is not especially dense in the region, but the gradual loss of the rainforest contributes increasingly to the problem of global warming and loss of biodiversity.

Unfortunately, many agricultural activities, in the rainforest and elsewhere, contribute to and are affected by climate change. Logging in the Amazon region of Brazil has been a long-standing contributor by releasing carbon into the atmosphere. Carbon that was formerly sequestered in tree limbs is released as the limbs are burned. Bovine flatulence, which is a by-product of cattle farming in the Amazon region and elsewhere, releases greenhouse gases that trap heat in the atmosphere. Decomposing rice-straw does so as well. These contributors to global climate change pale in comparison to the burning of coal and other fossil fuels, but are nevertheless significant.

Even in the best of circumstances, scientists figure that temperatures will rise in places such as India and Bangladesh by two degrees centigrade over the next 100 years. Such a rise may not seem like much, but it is enough to delay a monsoon, elevate the intensity of typhoons, and increase the frequency of droughts, as well as the severity of flooding.[36] Dry areas are expected to get drier, wet areas wetter, and storms more violent.[37]

Flooding is a frequent occurrence in Bangladesh, but in some years its people are literally up the creek without a paddle. I spent a few weeks in the summer of 1998 interviewing farmers during the worst flood in that country in recorded history. Water covered two-thirds of the country, in some areas for two months. It swallowed up houses, farm animals, and crops. You might ask why I bothered farmers at such a difficult time, but some of them did not have much else to do. Their crops were under water, and they were sitting on high spots—and a few roofs—waiting for the water to recede.

Unlike the people of New Orleans or New York, Bangladeshis expect some flooding almost every year, and many have prepared by building their houses on stilts. Rural roads are placed on ridges that have been built up higher than the fields. But years with deep floods inevitably mean suffering and sizable loss of life. Farmers suffer less than the landless workers who lose employment and have insufficient reserves to carry them through. If a flood coincides with a typhoon, as is common, squatters in low-lying coastal areas perish. The numbers can be as great as the 150,000 who perished almost overnight in a 1992 typhoon and flood.

One farmer told me that he spread the risk by having part of his farm in an "upland" area. Looking around at the water world that surrounded us for miles, I asked where the uplands were. He pointed to a small patch of land a short distance away. That patch was flooded as well, but it had an elevation of 12 feet above sea level. Uplands and lowlands are relative terms and have different meanings for people in Bangladesh versus Nepal. In the upland patch, the floods do recede more quickly, enabling the farmer to plant his rice again sooner.

As temperatures warm and oceans rise over the next few years, countries like Bangladesh will confront a host of challenges. More and more land will disappear under the waves. The soil will become increasingly salty, reducing its suitability for agriculture. With one-fifth of Bangladesh less than one meter above sea level and the oceans predicted to rise about two meters over the next century, the current uplands will become lowlands. Bangladeshis will have to make major changes in the way they farm or millions of people will lose their livelihoods. They will have to grow new rice varieties that are more resistant to salinity and periods of total submergence, abandon some of the lowest lying areas, and perhaps construct seawalls in some areas.

Sound management of natural resources is necessary for sustainable agricultural and economic development. The poorest countries are greatly dependent on their natural resource base because they are heavily agricultural, and thus most vulnerable to environmental problems. The livelihoods of many of the poor are natural resource based, and poverty causes people to place heavy emphasis on obtaining current outputs and income at whatever cost, including environmental degradation, in order to survive. Poverty also means more children, and the resulting larger population places added pressures on the environment. Institutional arrangements that might help protect the environment become outmoded as populations grow and societies become less personal. As a result, soils erode or become less fertile or saline, forests shrink, and soil, water, and air become increasingly polluted with chemicals. The expected climate changes will only compound these problems. Raising incomes and revising rules to reduce incentives for environmental degradation are essential for reversing these trends.

Growth in farm productivity and incomes reduces some of the environmental problems associated with agriculture. Facilitated by overall farm productivity growth, government programs, and lagging profits on marginal lands, farmlands with steep slopes, thin soils, and wet areas have been exiting US agriculture for decades. One result is that from 1982 to 2007 alone, soil erosion declined by more than 40 percent in the United States.[38]

As a child, I accompanied my father as we measured the areas in our fields to "set aside" from crop production so our farm would qualify for government payments on the corn and oats that we grew on the remaining land. The government program was intended to raise crop prices by reducing overall production. It seemed as though we were trying to trick the government as we set aside the areas that were marginal for growing crops and then applied extra fertilizer to the rest. Our overall production actually rose, and there was clearly an environmental benefit. The marginal areas we set aside at first became pasture and eventually grew back to brush and trees. The lands were steep or wet and should not have been farmed anyway. Farmers across America followed the same script, and millions of marginal acres grew back into the woods, wetlands, and wildlife areas for which they were better suited than farming. Subsequent productivity growth and farm programs have continued the trend to remove marginal lands and increase wildlife habitats.

However, I also witnessed mechanical cultivation for weed control being replaced by herbicide use on our corn. A pesky alfalfa weevil required insecticide applications as well. All across America, pesticide and fertilizer use grew. We are still struggling in the United States to wean our crops off dangerous pesticides and to keep excess nitrates out of our water sources—the latter from manures as well as chemical fertilizers.

As agricultural productivity improves in developing countries, farmers face similar environmental opportunities and challenges. Currently about 1.3 billion people obtain their living by farming fragile lands.[39] Populations living on these lands account for many of the people in extreme poverty. Increased use of farming practices such as reduced soil tillage and maintaining ground cover with crops and crop residues can help minimize erosion and store carbon in the soil. But in the long run, increased income, off-farm job growth, and reduced pressure to intensively cultivate marginal areas will create the primary opportunity for environmental improvement. Unfortunately, pressures to apply chemicals and irrigate with scarce water on the more productive lands are likely to continue, creating other environmental threats.

6

It Takes a Farm

When I was two years old, I spotted a newborn calf being nursed by its mother in the pasture behind our house. A curious kid, I rushed into the field to welcome the cute creature into the world. The cow was overprotective of her new offspring and charged, smashing me in the face and knocking me head over heels. That first memory in my life taught me that agriculture can be both beautiful and beastly—and dispelled any notion that I might ever want to be a bullfighter.

Farming sustains us with food, fiber, and fuel, but a simple dry spell, a small insect, a disease, or even a bountiful harvest that drives down prices can unleash a veritable fury on full-time farmers, especially those living close to the margin. Markandaya writes about small farms in India: *"To those who live by the land there must always come times of hardship, of fear and of hunger, even as there are years of plenty. This is one of the truths of our existence as those who live by the land know; that sometimes we eat and sometimes we starve."*[40]

Recent multiyear droughts in Somalia, Kenya, and elsewhere in the Horn of Africa have brought this lesson home hard to farmers there. Risk is part of farming and leads to conservative (risk-averse) behavior, especially by small farmers who make up the majority of the rural poor in developing countries. Though such conservative behavior is typical of farms with limited resources, as is sharing among extended family and villagers, it is difficult to generalize about small farms because of their great diversity and changes over time. Despite some similarities, each of the hundreds of farms I have visited is unique due to differences in climate, soils, topography, market structure, culture, family structure, and other characteristics.

Silverio Aristizabal farms two acres on a steep slope in the Andean region of Colombia. Silverio is fortunate because he lives in the coffee zone, the band of land that lies between 4,000 to 5,500 feet in elevation above sea level where the climate is moderately warm with periodic rainfall year-round. He plants nearly an acre of coffee, but never more than that due to the wide swings in coffee prices every few years. Colombian coffee growers received $.50/lb in 1993, $1.00/lb in 1995, $.50/lb in 2003, $1.00/lb in 2007, and $2.00/lb in 2011. Coffee bushes require two to three years from the time the seeds are planted to come into production, making it impossible for Silverio to quickly adjust his production level to offset price swings from year to year. Instead, he diversifies what he sells by planting sugarcane and plantain. He also plants a field of maize and beans for home consumption, and keeps a couple of pigs and a few chickens, much as we did on the farm where I grew up. He also plants cassava, a root crop that is very dependable.

His neighbor, Arnaldo, has some pasture and keeps a milk cow as well, but Silverio's plot of land is too small for a cow. Silverio works six days a week from sunup to sundown tending to his crops with his sons while his wife and daughters care for the animals and do housework. But he owns his land and his children all attend school. He has only a fifth-grade education, but his kids will have more. On the weekend he travels to town to sell his coffee, brown sugar, and plantain, and he buys rice and occasionally meat, among other items. His family has enough to eat, and to soften the blow when coffee and sugar prices are both low, he buys sundries such as toilet paper and soap to sell to his neighbors out of a small display case in the front room in his house.

Another of Silverio's neighbors, Amado Botero, farms higher up the mountain where it is too cold to grow coffee. He is not as well off. Amado plants potatoes and peas, has a small pasture, and even owns two cows, but his income is less than half of that of Silverio. He worries that the fungicide he applies to control the late blight on his potatoes will make him sick, as it did once already when he neglected to wash the pail he used to mix the chemicals before using the pail in the house. He has trouble controlling an insect that is damaging his bagged potatoes in storage. But like Silverio, Amado feels fortunate, at least compared to his neighbor Alba Cardona. She, too, has potatoes and pasture, but is a widow who does not own her house, land, or cows. She must give half of her production as a rental payment to the owner, who lives in town.

Life has improved for Silverio, Arnaldo, and Amado in recent years, partly due to the rise in crop prices and increased demand for livestock products. Life has been harder for Alba. All these farms remind me of ones we visited in the Peace Corps.

One of our favorite rural areas (*veredas*) to visit in the Peace Corps was called La Esperanza (hope). Farmers there were especially friendly and eager to try new things. The area is located about an hour south of Pensilvania, including a short drive to an area called La Rioja followed by a hike down the mountain. I had already built several gardens and rabbit hutches in La Esperanza, when a farmer from there named Luis approached me on the street in Pensilvania one day and asked if I would help him construct a rabbit hutch. We fixed a date and by coincidence, one of his neighbors, Manuel, also stopped me in the street that day and asked me where the extension agent, Luis Carlos, was because he had a discoloration on the leaves of his coffee bushes. Unfortunately, Luis Carlos had recently transferred to another town and the new coffee extension agent would not arrive for another week. So I told him I would stop by his farm on my way to visit Luis.

I normally did not make coffee recommendations, but with Luis Carlos gone, I was the only "crop expert" left in the Coffee Federation office in Pensilvania. I always wanted to pretend to be a coffee expert, so on the way to visit Luis, I stopped to identify the problem. Manuel was very pleased when I told him that the purplish color on his coffee leaves was caused by a phosphorus deficiency and not by a disease. Additional fertilizer would help, and I recommended a formulation I had heard Luis Carlos suggest to other farmers. Most of Manuel's coffee bushes were Arabica type and were planted in typical fashion under plantain (cooking banana) trees for shade. However, a few of his bushes were in the sun and some of them had what appeared to be a common fungal disease.[41] I told Manuel that additional shade and fertilizer would probably help those bushes, but that he might need a fungicide as well. I suggested he discuss it with the new extension agent when he arrived.

I said goodbye to Manuel and hurried further down the path to Luis's farm. Luis was ready. He had purchased the necessary wire mesh for the hutch floor and had cut down several *guadua* (thick bamboo) trees that we would use to build the hutch. A raised wire mesh floor is one secret to raising healthy rabbits, as they stay clean. I had brought two 10-pound New Zealand white rabbits with me. Luis paid me for one, and promised to pay for the other by giving a rabbit to a neighbor out of his first batch of offspring.

We started the construction by cutting and sinking in the ground six *guadua* posts to support the hutch. All of the cutting was done by machete. We cut the three posts in back slightly shorter than the three in front so the roof would slant to the back. We made horizontal posts to support the floor and roof, and nailed the wire mesh in place for the floor. We then used machetes to slice several lengths of *guadua* into narrow slats that we nailed in place for the sides, and fashioned doors and feeders in front of the three cages that made up the hutch. By that time it was 2:00 PM, and we were famished. We broke for lunch, which consisted of a plate of rice with a fried egg on top, fried plantain slices, beans, cherry tomatoes, and sugar water. Being a guest, I ate at a small table that his wife, Odila, had set up just outside the kitchen under the overhang of the roof. Luis and Odila ate in the kitchen. While we ate, a half dozen chickens pecked around in the dooryard, a pig that was tied up next to the house munched on a variety of scraps, and their thin dog stared intently at my plate. Coffee beans dried in the sun on a sheet of tin roofing that lay on the ground.

After lunch we finished the hutch by splitting several sections of hollow *guadua* trunks down the middle with our machetes to fashion half-moon-shaped tiles. These tiles were laid on top of the hutch for a roof. Luis cut tall grass that he stuffed in the feeders and placed a water dish in each cage. We decided to house the two rabbits together in one of the cages so they could start a family right away. They seemed quite excited about their new relationship.

Before leaving, I checked on the garden I had helped Odila build six months earlier. It was doing well, although some of her tomato plants were affected by early blight (*Alternaria solani*), a common tomato disease. I told Odila that a fungicide could be applied, but it was too late for the current crop as it mostly worked as a preventative. It would be best to remove the plants, rotate in a different vegetable in their place, and plant the next tomato crop on raised beds on the other side of the garden. She showed me the excellent compost she had made from poultry and pig manure, old plantain parts, and pulp from coffee berries. Poultry manure and the pulp that remains after the beans are removed from coffee berries are very high in nitrogen and other plant nutrients. Some parts of the plantain plant, especially the hanging spine-like shaft from which the flowers and fruit protrude, contain a natural biopesticide that can reduce a common soil-borne disease.[42] Experienced farmers like Luis and Odila are well aware of the fertilizer and biopesticide properties of such plant materials. They make maximum use of their resources, are efficient in their farming practices, and seek advice that might improve their situations.

Alam Karim farms one acre of land in Bogra, Bangladesh. His farm is flat and split into two parcels on which he plants two rice crops and one vegetable crop each year. The first rice crop grows in the spring, before the heavy rains begin. The second grows during the main monsoon season from July to October. One of Alam's parcels is in an "upland" area, and the other is lowland, which floods to some extent almost every August. By November, the waters have retreated back from the floodplains of the Brahmaputra River and what was water becomes land. Alam grows eggplant and a few pumpkins, beans, tomatoes, and cabbages during the dry and cool winter season from December to March.

Alam lives with his wife and four children. Because his children are young, he hires a laborer for 120 taka per day (about $1.50) during critical weeding and harvesting periods. Once every five to 10 years, he loses his rice crop to excessive flooding. In a normal year, his rice yields about 1.5 tons, which is enough to feed his family and to sell some as well. Most of his income is derived from selling vegetables, although prices are quite variable, and he loses a lot of his eggplant crop to a borer insect that causes the fruits to rot. Both the natural (floods, insects) and the supernatural (Allah) play a major role in his life.

Alam is not considered among the poorest of the poor. He owes some money to a local lender, but not much, and is able to visit a health center when he is sick. His two older children attend school. The laborer who works for him has no assets, and when floods or droughts come, he has no work. The laborer is lean, his body shape carved by work and hunger.

Alam worries because two of his children are girls. He loves them, but he must save for dowries if he hopes to marry them off to men of comparable age and means. Their future and his own depend on it.

Accompanied by my friends Ed, an entomologist (bug man), and Sally, a plant pathologist (disease woman), I visit a small farm in Nueva Ecija province in the Philippines to discuss methods a farm couple is using to manage crop pests. Gloria and Cezar Aragon farm 2.5 acres, planting rice in the rainy season from June to September and vegetables during the dry season from December to May. Their money-making crops are onions, eggplants, and peppers.

The whole family is busy harvesting onions when I arrive. Most of the onions are small red ones for local consumption, but a few are large yellow ones destined for the Manila market. Gloria and Cezar spread their risk by contracting early in the season to sell the yellow onions to a buyer who will stop by in his pickup truck to collect the produce after harvest. They do not receive a good price by contracting ahead, but locking in the price ensured a small profit so they could obtain a loan to plant, fertilize, and pay for help with weeding. A bird in hand is worth two in the bush.

Gloria keeps track of the money. She makes decisions jointly with her husband, a situation that is atypical in some countries, but not uncommon in the Philippines. She is responsible for the tethered goat grazing the dense, noxious weeds on the edge of the field. Her six-year-old son wields a sharp knife with amazing dexterity as he cuts the tops off the onions that have just been pulled from the dark soil. Gloria places the onions in a large netted sack.

The little kid with the knife reminds me of how dangerous farming is for children around the world. Kids start working at an age when their judgment is still suspect. But in fact, a six-year-old Filipino with a sharp knife is a lot safer than many US farm kids who work with complicated machinery. A 13-year-old who lived near us when I was a child got off a tractor attached to a manure spreader that was still churning away behind him. His jacket caught on a spinning part called the "power-take-off," which runs between the tractor and the spreader. It wrapped him up and took his life.

I had a mishap when I was 11 that thankfully had a better ending. I was stacking bales of hay on a wagon pulled by a tractor my great uncle Ernie was driving. A second wagon, already loaded with two tons of hay, was being drawn behind the first. My uncle stopped it every few seconds as my father, grandfather, and another uncle threw 40-pound bales onto the wagon. Uncle Ernie had arthritis that caused him to jerk the clutch so that the wagons periodically lurched forward and backward as we moved along. A hard lurch threw me off balance. I fell from the top of the load of hay off the back of the wagon, hitting the ground between the two wagons. The fall didn't hurt—an advantage of being 11—but I quickly saw there was little clearance under the second wagon. I scrambled along the ground trying to escape before being crushed by the wagon. I barely made it—the front tire of the second wagon rolled up to a stop about an inch from my face. My father had yelled at my uncle to stop the tractor, but Uncle Ernie was nearly deaf. He just happened to stop at a fortuitous moment.

I crawled out from under the wagon, and my father asked if I was hurt. I said no, but promptly began to cry. I knew I had just had the closest of close calls. To this day, I can see that semi-bald tire. Uncle Ernie was never asked to drive a tractor again.

Back in the Philippines, farm kids in Nueva Ecija do many tasks. They weed. They pick bugs off crops. They harvest. They bathe the water buffalo with muddy water to cool it down after it works in the field. Life on the farm is never boring for kids, and at a young age they know they are doing something that helps their family.

Ed and Sally ask Gloria what her main pest problems are. She points to some tiny red insects picking at her hands while she harvests onions and to some little tunnels on the leaves. She has sprayed an insecticide to control the two pests but yields are still down. Ed tells her the little red insect (called Thrips) is a nuisance, but not likely to be hurting production very much. The insecticide is destroying the good insects that keep the tunneling insect, a leaf miner, under control. Sally pulls an onion out of the soft ground and finds its roots are pinkish with little white balls attached to them. She suspects two problems, microscopic pests called nematodes and a disease called pink root. These are likely the major causes of the onion yield loss.

Gloria's pest problem illustrates several points that I first learned on my family's farm and later many times on farms around the world. Farmers know a lot, and if you want to know what they perceive to be their main problems, talk to them. But farmers do not know everything, especially about things they can't easily see, like nematodes and pink root. Yet they know they do not know everything and are generally eager for advice they think will help them.

Traditional farmers have knowledge that has been passed down for generations. They understand how farming practices relate to the environment. Many farmers grow beans and maize together because they know the combination improves their diets as well as the soil. Their diets are improved because beans add protein to the carbohydrates from the maize. The beans fix nitrogen in the soil, which benefits the maize, while the latter provides a stalk for the beans to climb. Traditional farmers diversify their crops and livestock to reduce risk. They know that sprinkling ashes on plants can keep certain types of ants away.[43] They understand little things about the weather. When I was young, weather reports were often inaccurate so we relied on numerous other indicators to predict rain. We would keep baling hay into the night if the dew failed to settle on the ground because that meant it would likely rain, and rain would wash nutrients from any hay left on the ground. We did not mow hay if the wind blew from the east or if a cloud bank resembled a furrowed field because those were sure signs of rain by the next day.

Despite their wealth of knowledge, traditional farmers are gradually forced to modernize as their country develops. They must adopt new practices that raise their land and labor productivity to compete in a highly competitive market and to meet the growing demand for food and fuel. The majority of farms around the world remain family owned and operated, but they grow in size over time to earn higher incomes for their owners. Or they become part-time farms to accommodate off-farm employment that can increase and diversify farmers' incomes.

I travel to Senegal with one of my graduate students, Ebere, to initiate a survey of peanut growers. Peanuts—also called groundnuts in Africa—were introduced into West Africa hundreds of years ago by Portuguese traders and have been an important crop there ever since. It is the dry season, and we travel by road from Dakar, the capital of Senegal, to Kaolack, a small city located in the groundnut basin. Dakar is a bustling city with a colonial French flavor. Some people call it the Paris of Africa—but not very many. Kaolack, on the other hand, feels a bit like a northwest Texas town in summer—only hotter, drier, and dustier. The air is dry and parches my mouth as we speed down the road. Our vehicle swerves to avoid trucks piled high and wide with bundled peanut straw destined for animal feed near Dakar.

In Kaolack, Ebere and I work with a local agricultural scientist named Matar to refine our survey questionnaire. We then visit some farmers to test the questions before using them in interviews with a large number of farmers. Leaving the single-lane road a few miles outside of the city, we drive down a dirt path and enter a farm "compound," a group of small circular mud huts with thatched roofs. White and black goats roam the compound and graze on dry peanut straw near one of the huts. We meet with an elderly man named Ahmadou, the family head, and his three sons. Each son lives in a hut within the compound with his spouse and children. Millet fields are scattered around near their huts, and several larger nonadjoining peanut plots are a short distance away.

Ahmadou and his sons describe their production practices and list their inputs, outputs, prices, and off-farm activities. Everyone defers to Ahmadou, who is the patriarch of this large extended family. Power is passed from father to son and brother to brother, based on age, in this traditional farm family. Food crops such as millet are grown near the compound to make use of any animal manure. The peanut crop receives little if any manure for fertilizer, as the amount is insufficient. Since the government program that subsidized chemical fertilizer ended, they have stopped using that as well. Peanut plants have nitrogen rich foliage, but after the harvest the dry foliage is collected and fed to livestock or tied up into bales and sold rather than being used as fertilizer for the next crop. As a result, peanut yields have dropped over time, along with soil fertility. Land used to be abundant, and Ahmadou would rotate fields, leaving some fields without a crop for years at a time so they could recover nutrients. Population pressure, from both his family and his neighbors, has gradually disrupted that system.

Ahmadou's sons say they earn some income by working part time in Kaolack. They do not make much, but when work is available, the extra income is helpful. The women and children do much of the work in the fields. The women are also responsible for domestic chores and for educating the children.

The family's biggest complaint is the high price of fertilizer and of other purchased inputs, especially since the value of the local currency dropped by half the previous year. The price of peanuts is almost never high enough to earn sufficient income for the family to eat well, so they need the supplemental income earned in off-farm work. Erratic rainfall is another major concern. With only one rainy season per year, if the rains fail one season, it is a long hungry wait for the next.

Despite the many differences in family structure and crops, I see similarities to our family farm in Hillsboro. We did not live in a compound—or in mud huts—but my grandparents lived in one house on the farm and we lived in another. Our extended family of uncles and cousins lived nearby and contributed seasonal labor. My father deferred many decisions to my grandfather. My mother played a key role in our education. Much of what we ate came from a large garden and orchard near the house and from our livestock. We depended on the government-supported milk price, much as Ahmadou depends on a government peanut price.[44]

As time passed, specialization and off-farm income became increasingly important for our farm as the economics of agriculture in the United States changed. Ahmadou's farm is experiencing the front end of a similar change. His sons and grandchildren will be pulled and pushed from the farm to the factory or service sector as wages grow and farm income is squeezed. His sons may continue to work part-time off the farm, but contribute on the farm during critical planting and harvesting periods. In an attempt to improve efficiency, his farm will likely specialize to a greater extent, and perhaps consolidate with a neighboring farm as labor costs rise and mechanization increases. Increased labor productivity will mean fewer farmers will be needed, and growth in demand for nonfood products will stimulate employment in the nonfarm sector.

The stories above illustrate a tiny fraction of the diversity of small farm agriculture. Nomadic livestock producers in Tanzania, wheat and sheep farmers in Syria, cotton growers in Mali, and communal maize farmers in Zimbabwe are among hundreds of other examples. Culture, climate, topography, government, history, religion, and many other factors combine to shape a vast array of distinct farm types. But as I visit small farms around the world, I am also struck by similarities: conservatism, hard work, importance of family, seasonal routines, efficient use of resources, multiple uses of livestock, mistrust of many government officials, concern over low prices for output, uneasiness with social change, a small margin for misfortune, guarded optimism, favoring the familiar but searching for ways to improve.

Despite its conservatism, the constantly changing character of small farm agriculture is also impressive. In the United States, the economic transformation that generated megafarms and retired marginal lands eliminated numerous small farms—and farmers—and turned most of the survivors into part-time operations. However, large numbers of small farms remain and continually reinvent themselves. In the 1970s and 1980s, many of these farms changed their product mixes so that on-farm labor demands complemented the hours of off-farm employment. Recently, rising incomes in the nonfarm population have spurred a demand for farms as entertainment—as evidenced by the spread of agrotourism in which people pay to visit farms, the growth in wine-tasting tours, and the proliferation of pick-your-own pumpkin, berry, apple, and flower farms.

Some small farms have gone organic; others have joined the local food movement.[45] As their incomes rise, a small but growing number of people are spending time and energy thinking about where and how their food is produced. Some of these people prefer to buy food from local farms with the hope that it is fresher, saves energy, and helps the local economy. They prefer products that are devoid of pesticides, antibiotics, and cross-species genetic modification. They purchase products from cows, pigs, and chickens that were raised in relative creature comfort. These products cost more than conventional foods, but their higher prices help some small farms to survive, at least until megafarms catch on and grab more of the markets as well.

In developing countries, many small farms will remain as economic development proceeds, even as farm consolidation occurs and nonfarm employment grows. A host of farmers will become part-timers, just as they did in the United States. The speed with which the transition occurs from mostly small full-time farms to a greater mix of farm sizes and types will depend on the existing size of the farm sector and the rate of growth of nonfarm jobs in each country.

Some countries have also struggled with the need to downsize historically large plantations or socialist farms, as discussed in part II, but the bigger future farm restructuring issues will involve farm consolidation, granting titles to land where only informal land rights existed previously, and the growth of part-time farming.

7

Trading Places

A woman sweeps the ground with a straw broom in a small dirt space near a traffic intersection in Manila where she is camped out with two little kids and a meager bundle of possessions. She sweeps as if she is in an imaginary house with invisible walls and someone has just spilled crumbs on the kitchen floor. It is better to have a home with an imaginary house than a house with an imaginary home, but I can't help wonder what unfortunate series of events has led her family to set up temporary residence at this particular crossroads. With every sweep of the broom she fights hard to make the best of a difficult situation, to provide a sense of normalcy for her family and a little dignity—all she has left of the nothing she has.

The woman just described most likely had recently migrated from a rural area to Manila. People move from rural to urban areas for many reasons, but by and large people move because they expect increased economic and educational opportunities and improved services. In some cases they are pushed out of rural areas by poverty or abuse, but usually they are pulled to cities by hope and opportunity. Migration within a nation is, in part, a natural reflection of the decline in relative size of agriculture as compared to industry and services as an economy grows. Farms consolidate and marginal lands fill with people until residents decide to move elsewhere in search of better jobs. When incomes grow, people want to consume more than food, so the jobs will be mostly off-farm, unless farm productivity lags so much that most people have to keep farming marginal areas just to survive.

Migrants tend to be young, single, and better educated than those left behind. They often have relatives or friends in the cities where they are headed. These characteristics reduce the cost of the move and raise the expected benefits. But when desperation drives people to move, perhaps after a natural calamity or because of domestic abuse, all bets are off.

When Marj and I lived in Colombia, we would take a bus about every three months from Pensilvania to Bogotá or Manizales. About three hours into the trip, the road came to a T-intersection where we could stay on the bus and go left to Bogotá, or we could transfer to another bus and go right to Manizales. Once when we approached that intersection, a woman with two small kids and a large bundle of goods asked the bus driver which city was closer, Bogotá or Manizales. He told her Manizales, at which point she and her children changed buses. It appeared to us they were moving from the rural area but did not even know where they were headed when they got on the bus in Pensilvania. They seemed to be desperately running away from something.

Most people who migrate from rural to urban areas do not arrange for a new job before they leave the rural area. Their interval of complete unemployment in the city tends to be short. They work at any informal job they can find to survive while looking for something better. When my car pulls up to a stoplight in most capital cities in low-income countries, someone offers to clean my windshield or sell me cigarettes, a wash cloth, popcorn, a plastic bag full of water, or a coat hanger. It is not much of a job, but a stint in street sales is a living, at least a temporary one. In some cases, people view the move to a city as a stepping stone, not only to a better job but to a possible move to a country such as the United States where they think most people are rich. They have hope.

Living conditions for new migrants in capital cities are often poor—shanties with limited or no services. The transition from a rural to urban area can be hard. Once during an economic downturn in Ecuador, I saw a family of indigenous people, apparently new arrivals and homeless, huddled together in the rain on a traffic island in Quito—hard indeed.

A few countries, such as China, attempt to regulate internal migration to encourage decentralized urban growth, but such restriction is impossible and probably undesirable for most countries. People desire freedom to seek out economic opportunity. The best means to stem overly rapid migration to capital cities would seem to be policies that encourage growth of jobs, services, and other amenities in smaller cities and towns.

Remittances, the money that migrants send back home, are the lifeblood of many families and an important source of foreign exchange for countries such as the Philippines, Honduras, Mexico, and Bangladesh. Remittances provide more capital than foreign aid for most developing countries—in Honduras it provides more than a fourth of the national income. Many workers on farms and in service jobs in the United States have come from other countries on temporary work visas. They migrate here for part of the year and send money home to their families. They have become an important source of US farm and nonfarm labor. Even more common are family members who move to the capital cities in their home countries and send much of their earnings to relatives living in farms and villages. In rural areas, many houses and lives have improved thanks to remittances from relatives who left for the city.

When I travel to Bangladesh, I often change planes in the Middle East in Kuwait or Dubai. Over 90 percent of the passengers on many flights into Bangladesh are male workers coming home on leave from jobs in the Middle East. In baggage claim, most of the pieces circling the carousel are bundles tied up with rope. Each worker has wrapped his travel needs in a bedroll, like a cowboy on a cattle drive. My bag is easy to spot. It is one of only a few lonely suitcases.

Female migrant workers are common on flights between the Philippines and countries like Singapore or Kuwait, which depend heavily on Filipina workers. They spend months each year as household helpers, thousands of miles from home, in attempts to better their own and their families' situations. Almost a fourth of the total Filipino labor force lives and works abroad.

Often lost in immigration debates are the effects of temporary workers who take jobs that citizens of the employing countries often refuse. In the United States, many of these workers are farm laborers from Mexico and Central America, hired especially at harvest time. Employing them keeps production costs down, which helps consumers and overall economic growth, although working conditions can be harsh. The money the temporary workers send home also stimulates growth there and creates markets for US goods.

House of Julio and Imelda Ospina (the children pictured are Mario, Cecilia, Estela, Marino, Fabio Alonzo, and Esneda).

Pensilvania, Caldas, Colombia.

Charles raking hay on the Norton farm.

Farmer building a rabbit hutch in Colombia.

Luis Carlos (extension agent) instructing farmers in Colombia.

This fish pond provides protein for a farm family in Colombia.

Marj and John on Sisseton Wahpeton Sioux tribal farm.

Breaking bricks in Bangladesh.

Houses in Dhaka,
Bangladesh, during
1998 flood.

Boy in the Philippines taking buffalo to a watering hole (Sally Miller).

Farmer spraying pesticides in the Philippines.

Farmer planting cowpeas in Ghana.

The "hungry season" is hard on children in eastern Uganda.

Typical rural bus
in Colombia.

Marj in Peace Corps.

Students interviewing farmer in Ecuador.

Clothing factory in Bangladesh.

Bangladesh farmer using IPM CRSP technology.

Progressive Ecuadorian farmer.

The Stories: Part II

⸺⬥⸺

Winning the World Food Fight

Colombian women seeking instruction
on how to vaccinate a chicken (Marjorie Norton).

8

Building Bridges

Before I visit certain countries, friends sometimes ask if I am worried about kidnapping or political unrest. The answer is no. They are much lower on my list of concerns than are roads. In the Andes, the foothills of the Himalayas, and numerous other rural areas, narrow roads are carved out of mountain sides with few or no guardrails. Even in countries where the roads are relatively flat, they are often narrow and rough. Roads and bridges that link farms to markets are vital to agricultural development. In too many developing countries, rural roads are rudimentary at best. They wash out or turn to mud during heavy rains. Poor road conditions increase transport costs by causing trucks to breakdown frequently. They reduce the prices farmers receive for their goods, as marketing agents factor in their high costs and the risks of product spoilage during transport.

And the risks are not only to goods but to lives as well, as accidents are common. For example, in Bangladesh, roads are flatter and smoother than in some countries, but they are narrow, and trucks and buses fly like bats out of—umm—Bangladesh. A road trip in Bangladesh can really get one's adrenaline flowing. Trucks, buses, cars, and bicycle rickshaws share single-lane roads with just two primary rules: (1) maximize speed and (2) when two vehicles approach each other, the bigger one has the right-of-way. If your vehicle and the one approaching it are roughly the same size, the tie is broken by a heart-stopping game of chicken. I suffer from motion sickness, and so during each trip I have to choose between nausea in the back seat or fear factor in the front—like riding shotgun in a race at Daytona.

One evening on a trip to Bangladesh, I chose nausea and was riding in the second row of a van. Ed, Sally, and other friends were with me. As we drove through an intersection, a large colorful truck appeared out of nowhere. It slammed into the left side of our vehicle, T-bone style. I flew across the seat in a shower of glass as the side of the van crumbled inward where I had been sitting. Our driver had violated rule number 2. We were fortunate. I suffered only a slight bruise and a minor cut. No one else was hurt.

A crowd of people quickly surrounded our mangled van, giving in to the universal urge to gawk. The police arrested the truck driver. Apparently there are limits to rule number 2.

In urban Asia, drivers are also schooled in beating competing vehicles to the slightest open space, except in Vietnam, which has its own unique (if unwritten) traffic rules. There, motorbikes are king, and riders are respectful of pedestrians. Streets in Ho Chi Minh City are always a sea of bikes, leaving the uninitiated pedestrian puzzling about how to cross the street. The secret is simply to step into the traffic and walk across the street at a steady pace. The flow of bikes parts around you like the Red Sea for the Israelites out of Egypt. But try that walk-off-the-curb trick in Lima, and you will be off to see your maker.

Modern infrastructure such as all-weather roads and bridges, rural electricity, water and irrigation systems, and high-speed communications are essential for moving farms from barely surviving to thriving. In the United States, the drop in transport costs due to the succession of improved gravel roads, in-land canals, railroads, paved roads, and interstate highways was partly responsible for the long-term growth in farm production. Electrification spread from about 10 percent of rural households in 1935 to almost complete coverage by 1960—a period of rapid farm productivity growth. Irrigation was essential in making California the largest producer of agricultural products in the United States today. Modern rural infrastructure does not guarantee agricultural development, but sustained farm productivity growth is difficult without it.

The danger on rural roads hit us personally many years ago in Colombia.[46] It was early on a hot afternoon, and I was helping a farmer start a vegetable garden when I began to feel queasy. I sat, rested for a few minutes, and went back to work. About a half hour later, stomach cramps signaled that the local sanitation had caught up with me. Marj was working nearby with a group of women and saw my plight. She encouraged me to return to town on our motorbike. She would take the bus when it passed by later, as she still had work to complete. I pondered what to do, but decided to stay and finish the garden.

About 6:00, shortly after the bus had passed by, Marj and I headed back to town on the motorbike. The sun was beginning to set, and the temperature was cooling down. With no traffic to worry about, I twisted the grip on the handlebar to increase speed and hasten our arrival before cramps grabbed me again. We rounded a curve and the sight of a small unexpected crowd on the side of the road caused us to quickly slow down and stop. Some people looked distraught. Others were scrambling down the steep mountain slope below the road. They were all focused on the little that was left of a bus hundreds of feet down, the one that Marj would have taken had I gone home earlier that afternoon on the bike. The bus had run off the road and rolled down the mountain, crushing passengers as it broke up. Many were killed. We felt fortunate, but exceedingly sad for the needless loss of life. Poor roads and drunken drivers are a toxic mix anywhere in the world, and unfortunately, all too common.

Poor road conditions can create severe economic losses. Marj and I spent a week in Pensilvania when no buses or trucks moved in or out of town after dozens of landslides covered sections of the road after a major storm. The economic losses were sizable for farmers needing to transport their products, especially perishable ones like fruits and vegetables.

Fortunately, significant improvements have been made in rural roads in many developing countries over the past few decades. Paved roads with drainage ditches are gradually replacing dirt roads in places like Ecuador, Colombia, India, and Ghana. The road where the bus accident occurred has now been improved and paved. Dirt roads are replacing footpaths in remote areas of countries like Nepal. It is a start.

Roads are but one of the infrastructure needs in rural areas of developing countries. Power, communications, and water systems are also critical for improving agriculture and reducing poverty. Rapid rural electrification has taken place in many developing countries in the past 40 years, but many parts of Africa and some other remote areas are still without power lines to houses that lie beyond villages. More than half of the people in sub-Saharan Africa do not have electricity, including more than three-fourths of those who live in rural areas. Rural residents with power often suffer from stoppages and brownouts. The lights become too dim for reading during hours of peak usage. In Nepal, even though power lines reach most areas, electricity is in short supply and it is regularly rationed. In rural Nigeria, The Gambia, Bangladesh, Ecuador, and many other countries, power outages are so frequent that farms or firms that can afford them rely heavily on backup generators. When power is erratic or unavailable, long-distance communication is difficult; cell phones and computers need recharging.

In some areas, the use of solar power is expanding. We recently conducted a farm household survey in eastern Uganda. Most of the households had no electricity, but a few had small solar panels that could light four bulbs and charge a cell phone. The panels cost about $55, a big investment for the poor, but their use will likely grow.

Despite power outages, one of the most helpful inventions for farmers, fishermen, and other entrepreneurs in recent years is the cell phone. Cell phones allow farmers to quickly access price information, enabling them to receive higher prices for their products. Price variation has declined as farmers can rapidly obtain information on maize or onion prices and are no longer at the mercy of traders with a monopoly on the latest information. Fishermen off the coast of India use cell phones to determine which port will offer the highest price for their catch each evening. Even poor rickshaw drivers in rural towns in Bangladesh use them to find customers. Cell phones are dirt cheap in developing countries. Nobody purchases a plan; they simply buy minutes. Cell phone connectivity has obviated the need for landline technology in many areas. And with cell phones comes the Internet in areas with electricity. The resulting improvements in communication provide a small measure of power to the poor.

Water infrastructure is critical for irrigation as well as for basic household consumption and health. I once took an undergraduate student, Alison, to Ecuador where we interviewed several farm families to study why the local health data showed significant variability in illness and death from diarrhea-related causes. We asked them questions about their water source and storage; livestock access to their water supply; existence, nature, and location of their latrine; whether they washed their hands with soap after using the latrine; occurrences of diarrhea in the past month; and many other questions that I would be uncomfortable answering, but they answered willingly. Alison, who was majoring in biology and chemistry at Virginia Tech and was soon to enter medical school at Tulane, also tested the water for assorted microscopic critters.

After interviewing several dozen families and analyzing the data, we found one factor that appeared to be related to incidence of diarrhea: livestock access to the water used in the house. In some cases, water came from streams where cattle drank or walked. In other cases, the water source was clean, but livestock contaminated it as it was piped or channeled through a pasture to the house. In a few cases, cattle or pigs could sneak a sip from the container that held the water near the house. Bacteria and parasites can build up in the body to the point where they cause weight loss and illness. I learned in the Peace Corps that one can have a bit of an iron stomach for a while, but eventually the little buggers cause problems if not purged from the digestive system. Some parasites are small but painful, and others are large and gross. A big health danger, especially for children, is the deadly dehydration that can follow diarrhea.

De-worming pills can help reduce the parasite problem, and low-cost chlorine drops in the drinking water can help eliminate bacteria. Where feasible, a more expensive but permanent solution is to isolate and enclose the water source and the pipes that carry water to the house. The appropriate solution for a particular situation depends on the resources available for improving water supply infrastructure, but de-worming pills and chlorine are cheap and underutilized in much of the developing world.

The Sisseton Wahpeton Sioux dodged a bullet. The tribe had planted hundreds of acres of corn, barley, flax, and sunflowers despite extremely low subsoil moisture. If the rains had failed to fall at just the right times that growing season, the tribe would have lost a bundle, just as they had the year before I was asked to advise them on what, where, and how much to plant, and how many cattle to run. The farm plan had looked so simple in my computer model, but as we watched the crops and waited for the rains in June and July, we held our breath. I spent some sleepless nights worrying whether Mother Nature would be cruel to the tribe once again. Fortunately it did rain like clockwork almost every week that summer. The harvest was heavy with grain and the pastures were flush with feed.

But that fall, as I programmed the tribal farm model again to evaluate long-run tribal farm investments, I examined the feasibility of introducing irrigation. Even in a normal year, rainfall on the reservation averaged only 22 inches per year, less than half the amount we were accustomed to on our farm in upstate New York. And the rainfall amounts are highly variable in South Dakota. Crop yields could easily double under irrigation in the plains and the certainty of having adequate corn and hay would reduce the risk associated with feeding cattle. The wells and irrigation equipment were expensive, but with financial assistance from the federal Economic Development Administration, the tribe was able to invest in irrigation for several of its fields, which improved the profitability of the farm for years to come.

Farmers and government agencies around the world struggle to decide how much emphasis to place on irrigation investments. Most high-yielding seed varieties require complementary applications of water and fertilizer if they are indeed to yield more than traditional seeds. The added production is needed to meet the growing demand for food, yet water uses often conflict in developing countries, just as they do in California, where farmers and city dwellers compete for scarce water supplies. Water availability and allocation have become increasingly important concerns as population and agricultural production continue to expand.

Governments must also decide whether to support large-scale dams and irrigation channels or to emphasize small-scale irrigation schemes. Water use efficiency is critical, and large-scale systems that use surface water are often less efficient than smaller ones, although groundwater is limited in many locations and requires energy to pump it. It is important for available water to be priced at its true value in alternative uses. Government programs that distort water pricing can create major inefficiencies in water use. Some crops, such as rice, are heavy water users and often rely on fields being flooded by seasonal rainfall. Few issues will be as important to agriculture as efficient water use over the next several years.

9

Seeds of Hope

In much of the developing world, reducing poverty means improving agricultural productivity. Even the poorest of farmers, although they favor the familiar, continually seek ways to better their situations. Given their existing resource bases and technologies, farmers are pretty good at what they do. But in many cases, current technologies only provide low yields, and soil nutrients are extracted faster than they are replenished. Given the difficulty of expanding their land base, farmers need something else, such as higher yielding plants and animals, pest-resistant and drought-tolerant plants, improved fertilizers, and more nutritious plants. But new technologies require resources and involve change, not all of which is positive for everyone. Trade-offs must be carefully considered.

Millions of farmers, especially in Asia and Latin America, faced a dilemma. They wanted to raise their wheat and rice yields above the lousy levels they were experiencing, but they took a huge risk when they increased the water and fertilizer they applied to their traditional rice and wheat plants. The extra fertilizer and water would increase the leaf mass, causing the crop to collapse under the extra weight during the first typhoon, strong monsoon, or El Niño that hit. Their crop would fall flat in the driving wind and rain.

A scientist named Norman Borlaug noticed this problem and, with a cadre of other agricultural scientists, bred new wheat and rice plants that had shorter and sturdier stems. Applying extra water and fertilizer to the plants led them to produce more grain with less foliage. These new plants were wildly popular among both rich and poor farmers, especially among those with access to irrigation. The result was that millions of the poor beat the odds and now live and eat better.

Bur crop breeding is not for the impatient. Borlaug began his breeding program with wheat in Mexico in the 1940s, and it was not until the 1960s that the first new plant varieties were released. By the 1950s, it was clear that he was onto something big, and other scientists working in the Philippines took the same approach and developed new high-yielding rice varieties.

In the mid-1960s, India and other countries in South Asia were experiencing serious shortfalls in food production. The population growth rate in the world was at an all-time high, and some pundits were predicting global starvation by the 1980s. The short-straw rice and wheat arrived just in time. Millions of people were saved from starvation and a "green revolution" occurred that bought the world time. This time "short straw" meant luck rather than loss.

The green revolution varieties still have their detractors. They were accompanied by increased use of fertilizer and pesticides. Although subsequent work by breeders has built resistance for many insect pests and diseases into the plants, an expanded market for chemicals became established that has been difficult to counter. Farmers have come to see pesticides as "plant medicine" and seem to be addicted to their use. Efforts are underway to change their mind-set and find new ways to increase yields, but it has not been easy. A network of 15 international agricultural research centers[47] around the world is engaged in a continual struggle, in partnership with research institutions in developing countries, to develop sustainable agricultural technologies. However, the green revolution has run its course, yield growth has slowed, and food prices have begun to rise. Very little new land can be brought into production, the world population is projected to reach 9 billion by 2040, foodstuffs increasingly compete with biofuels for farm resources, and water is increasingly scarce. New approaches are needed to raise agricultural productivity, especially in Africa. The approaches must be environmentally, economically, and politically sustainable.

When I arrived in the airport in Owerri in southeast Nigeria with my graduate student, Nderim, we were met by two men in an old pickup. They drove us the hour to Umudike, where the National Root Crops Research Center is located. One of the men was a scientist, the other a bodyguard holding a shotgun. Robberies and kidnapping are common in

Nigeria, especially in the southeast, and with foreigners in the vehicle, they came prepared. We drove through a humid area abundant with plantain, trees, cassava, and churches.

We were on this trip to meet with two plant breeders, Emmanuel Okogbenin and Chiedozie Egesi, who were working to improve the production and nutrition of cassava, a crop that is literally the savior of many poor farmers. Nigeria produces about a quarter ton of cassava roots per person per year. Cassava is consumed there as a sticky ball of cooked starch called *gari*. The crop is mostly carbohydrates, but is favored by the poor because it grows in wet or dry conditions, tolerates poor soils, and has a long harvest season. It can fill stomachs while farmers await the next grain harvest. It is like money in the bank.

Unfortunately, cassava also has few vitamins and minerals, is susceptible to insects and diseases, and does not store long in fresh form. It must be consumed or processed within a couple of days or it turns black and rots like an overripe banana. I have worked on two research projects that have been successful in reducing cassava pests, extending its shelf life, and fortifying it with vitamins and minerals. My part of the research involved assessing the net benefits of modifying cassava in this way, while biological scientists did the modifications.

Scientists such as Emmanuel and Chiedozie use molecular means to speed up their breeding work. One approach is to sleuth around in Brazil, the cradle of cassava, to identify wild species of the crop that may not be edible but are resistant to pests or have a longer shelf life than domesticated cassava. The plants are taken back to the lab and the scientists identify the genes in the cassava plant that control those traits and transfer them to an edible variety by "crossing" the wild and the edible varieties. Then, unlike in traditional breeding, they study the new plant at a molecular level to make sure the improved trait was indeed transferred. This method is called marker-assisted selection, or MAS.

Another approach is to transfer the genes directly rather than just cross the varieties. This approach, which is called genetic engineering, works better than MAS for fortifying the plant with vitamins and minerals because the genes for those traits may not come from wild varieties. However, genetically modified organisms (GMOs) are controversial because of fears about unexpected consequences.

Possible consequences of GMOs have been studied and few negative ones of significance have been found. Countries have implemented rules and tests that must be conducted before approval is granted for release of a "genetically modified" plant. But this GMO approach has been less popular in developing countries than in developed countries, at least so far, due to fears, regulatory issues, and vested interests. Near Umudike, we visited with women who were cultivating cassava in their fields. They had heard of the new cassava plants that would reduce susceptibility to diseases and insects and wanted to know when and where they could get them. We said, "Be patient. They are coming—God willing." I think they appreciated the religious touch as they started chanting, dancing, and saying grace as we broke out juice boxes and chips.

Half a world away, scientists in the Philippines and Bangladesh used MAS to modify popular rice varieties to withstand a lengthy period underwater during a flood without drowning the plants. They also used MAS to develop rice that can tolerate salt water. These new plants should help poor farmers adapt if global warming leads to increased coastal flooding. Strides are being made, but there is still a long way to go in developing improved varieties of basic staple crops, particularly because of climate change and the way insects and diseases keep evolving as they attempt to outsmart the scientists.

About five years into my research program at Virginia Tech I was asked by the national agricultural research systems in Ecuador and the Dominican Republic to help them design and implement a process for establishing research priorities. Research resources in these countries were scarce, and they had many objectives they hoped to achieve, including improvements in farm efficiency, increased incomes for small farms, and food security for consumers. I helped them to develop a framework for combining data on production and markets with expert opinions on potential changes associated with alternative crop and livestock technologies. The framework allowed them to explore the implications of emphasizing alternative objectives.

Due to that effort, the International Service for National Agricultural Research (ISNAR) asked me to assist other research systems in Asia, Africa, and Latin America with priority setting. ISNAR provided funds to assist the systems in a dozen countries and to prepare written materials.[48] One of those systems was in Zimbabwe. When I first visited there in 1989, we held a priority-setting workshop with scientists and administrators. Zimbabwe was struggling with a dual structure in its agriculture. It had about 3,000 large commercial farms owned mostly by white farmers and 600,000 very small farms owned by black farmers. Attempts to reform the land owner-ship pattern were moving slowly, and one of the issues was how much to support each type of farm. The large farms were productive and feeding much of the country, while the small farms were impoverished.

At one point during the workshop, a scientist stood up and argued that we were wasting our time because, in the end, witchcraft determines when and where research succeeds. This individual appeared to sincerely believe what she said, and her views made me wonder if reforming the research system would be even more of a challenge than I had thought. I thanked the speaker for her comments and we moved on in the discus-sion, as others did not follow up on her remarks. During a break in the meeting, several scientists came up to me quietly and told me not to be concerned about her comments as they represented an isolated view among themselves.

A couple of years later, the Rockefeller Foundation asked me to again assist with agricultural research priority setting in Zimbabwe. Some issues were unresolved and the new director of research had requested assis-tance. Ironically, the new director was the same person who had made the witchcraft comment. As we talked, I could sense that she sincerely wanted to improve efficiency in the research system and to help impoverished farmers. She said she liked my approach two years earlier and did not mention witchcraft. The weight of her new position may have nudged her away from her magical view of the world. I doubt my scintillating presen-tation on research priority setting methods caused a spiritual conversion.

Assisted by one of my graduate students, Gladys, I agreed to work with the director and other research leaders from Zimbabwe to conduct an analysis aimed at determining the set of research activities that would provide the greatest gains in agricultural productivity, and then to assess how much would be sacrificed overall if the system focused heavily on technologies aimed at small farms. It turned out not that much would be given up, which pleased the director. Still, I wondered how she had risen to the top in the system. It is not unusual to be surprised by who secures an appointment, as directorships around the world are often political positions. But the previously stated views of the director in Zimbabwe had been out of the ordinary for a scientist, and they reminded me of the young earth creationist views of a tiny minority of US scientists. I found out that she is a sister of President Mugabe. To her credit, she performed relatively well until her aging brother and his followers made a mess.

I have met directors more unusual than the one in Zimbabwe, and fortunately many excellent ones. They and the agricultural scientists under them work hard—even though most are poorly paid. Most researchers are dedicated individuals who make a difference in the lives of the poor. They are also eager to interact with other researchers at international centers and universities around the world. The research systems vary in size from a handful of scientists and two research stations in The Gambia to the Chinese system, which has more agricultural scientists and stations than the United States.

When I first came to Virginia Tech in 1980, I inherited a research project that was funded by the US Environmental Protection Agency. The purpose of the research was to reduce crop pests while also reducing the use of chemicals harmful to people and the environment. The project had an entomologist named Ed (the same Ed mentioned in earlier stories), a pathologist named Ernie (not my uncle), and me. I enjoyed working with scientists from other fields and learned a lot about plant pests, pesticides, and new ways for farmers to manage pests safely.

A few years later, the US Agency for International Development issued a request for proposals for an Integrated Pest Management Collaborative Research Support Program (IPM CRSP). They would fund a global project aimed at improving farm productivity while minimizing chemical use, especially on vegetables, in developing countries. IPM does not necessarily rule out the use of all chemicals, as organic farming would, but it minimizes their use through alternative practices such as biological pest controls and cultural practices. The project sounded like it was up my alley, given my interest and experience in developing countries, knowledge of pest management, and my domestic and international contacts. It would provide several million dollars for research and education, and would involve more than 100 scientists in the United States, Latin America, Asia, and Africa. Projects like this usually involve scientists from a group of US universities who collaborate with each other and with scientists in research institutions in developing countries.

I discussed the request for proposals with our new director of international programs at Virginia Tech, S. K. De Datta. He called his counterpart at a university that we knew would be one of the favorites to win the proposal competition. S. K. asked if we could join their team, but was turned down because they already had enough partners. Before coming to Virginia Tech, S. K. had worked for almost three decades at the International Rice Research Institute (IRRI) in the Philippines and was among the group of scientists who produced the green revolution in rice. He was confident Virginia Tech could lead the project, and suggested we assemble our own team. We asked scientists from Penn State, Purdue, Ohio State, and other institutions to join us. I agreed to write the proposal and a few weeks later we won the project. Our formula for success had been the interdisciplinary and participatory nature of our approach. It linked scientists to farmers in Africa, Asia, and Latin America.

That was 1993. Since then, the IPM CRSP has developed IPM programs in more than 20 countries in the developing world, and the program continues to grow. Many thousands of farmers in the Philippines, Bangladesh, Indonesia, Ecuador, Honduras, Mali, Uganda, Albania, and elsewhere have adopted IPM practices that have reduced pest problems, minimized pesticide use, and raised farmers' incomes. US scientists meet with developing country scientists to plan, implement, and review progress on a regular basis. Much of the research occurs in farmers' fields, and much of the IPM technologies and information are spread to farmers with the help of local government, nongovernmental organizations, as well as the private sector.[49] More than 150 graduate students have completed degrees on the IPM CRSP. Many of them have returned to their home countries to continue the work alongside the hundreds of other local scientists who have been involved over the years.

Ed, Sally, and I often visit Bangladesh on IPM CRSP work in late January because it is the peak period for growing vegetable crops. We meet with scientists and farmers. One day we walked onto a farm where the farmer was using an IPM practice that we developed to reduce a problem with fruit flies on his one-acre pumpkin field. His crop looked good, and we asked if he was happy with the results. He had been using the IPM practice for two years. He pointed to his block house with its tin roof. He said that the additional money he made had allowed him to replace his previous mud house and thatched roof. He had become almost middle class by Bangladesh standards.

Later, Ed, Sally, and I attended a farmer field day designed to spread information on IPM. Growing up on our farm at home, I used to attend demonstrations organized by county extension agents that were designed to spread information about the latest farm practices. We were lucky if 50 people were there. This day in Bangladesh, more than 1,000 farmers showed up. Of course, when farms are only an acre in size, 1,000 people in a densely populated area can come from a small radius. I suppose the fact that we gave out lunch may have helped participation as well—like the pizza trick that I sometimes use to attract college students to an event. The Bangladeshi farmers moved in groups around the fields, observing, listening, and asking questions about the new vegetable technologies.

Wherever you are in the world, a gathering of 1,000 people will usually attract politicians, especially when something positive has occurred and credit can be claimed. Politicians love to give speeches. We encourage them because support from political leaders can help reinforce the IPM message.

After touring the fields, the large group of people gathered around a stage that was set up nearby. When the Bangladeshi scientists and politicians, Ed, Sally, and I approached the stage, the crowd divided to let us through. As we walked through the crowd, kids moved forward and sang and tossed flower petals at us. It reminded me a little of a wedding and of the football team entering the stadium at Virginia Tech between rows of cheerleaders and students before a game. The Bangladeshis did not play "Enter Sandman" as they do at Virginia Tech, but the flowers were a nice touch. I couldn't help but reflect on the fact that we were Americans being greeted warmly in a Muslim country while a war raged two countries away in Afghanistan.

The scientists, politicians, some farm leaders, and I spoke to the crowd about IPM. Then three farmers came forward. They played instruments and sang an original song they had written about IPM. Ed, Sally, and I felt a little like judges on "Bangladesh Has Talent." Bangladeshis know how to show appreciation.

Of course, not everyone is as tuned in to the IPM CRSP, or at least to the involvement of US scientists. Ed, Sally, and I visited farms in South India that were using IPM CRSP practices suggested to them by researchers from the local agricultural university. While we were there, the researchers gathered a group of farmers to speak with us. We shook hands with them and sat in front of a farm house where the women served us tea. The farmers spoke Tamil so the researchers spoke to them first and then translated as we asked a variety of questions about their pest problems and management practices. The farmers dutifully answered each query, and near the end I asked if they had any questions for us. A woman in the group stood up and asked, "Who are you people, and why are you here?" We laughed as we realized we had entered their home like long-lost relatives and they had not a clue as to who we were. Their prior contact had all been with Indian project scientists. Next time we would be more careful to introduce ourselves properly!

R. Muniappan (Muni), the current director of the IPM CRSP at Virginia Tech, like most entomologists, is fascinated by insects and always carries around a net. On a research trip in 2008 to South India, he walked up to me all agitated. He had just captured a papaya mealybug, or PMB. He explained that the PMB was first reported in Mexico in 1955. It did little damage to crops there because natural biological enemies kept it under control. But when it moved to other countries in the Caribbean and Southeast Asia, it caused significant losses in papaya and other tropical fruits, vegetables, and cassava. Until the day Muni spotted it, the PMB had not been reported in India.

For the next two years, Muni was a man on a mission. He pressured the Indian and US governments for permission to import a small parasitic insect into India to implement a biological control program for PMB. However, before permission was granted, the PMB population exploded and decimated the papaya crop, then began to attack several other crops in the southern part of the country. Fortunately, thanks in part to the efforts of Muni, the parasitic insect was then introduced and released. Almost overnight, the PMB populations dropped and the crops bounced back. When we visited again early the next year, farmers were ecstatic. Government officials gave speeches extolling the parasite release program and Muni for saving the crops. His alertness, persistence, and cooperation with Indian scientists and officials meant that more than $100 million in crop losses were averted the first year alone, and the pest did not spread to major agricultural areas in northern India.

Speaking of losses averted, while Marj and I were in the Peace Corps, we were dining with a friend in his home one evening when we heard what sounded like rats scurrying around overhead. We looked up and saw a large jagged hole in the ceiling. I asked our friend if he ever had a problem with rats falling through the hole. He replied no, never. The words were barely out of his mouth when the largest rat we had ever seen fell through the hole and landed with a thud, right in the middle of the table. Marj jumped up on her chair and screamed. We chased the rat around the room with a stick and an umbrella until the racing rodent bounded out the back door into the garden, never to be seen again—that night.

Marj and I periodically battled rats in our own house while in the Peace Corps. We placed poison in a few places to attract and control them. We were sleeping one night when something, which felt like an animal, crawled across my chest. I quickly brushed it off and turned on the light only to see a cockroach the size of a mouse speed out of sight, with many of its relatives in tow. Relieved, we turned out the light and went back to sleep. What are a few roaches when you are worried about rats?

Rodents are a continual problem in most developing countries due to the prevalence of accessible garbage, tropical rice fields, and unsealed storage bins for grain. They carry disease and can consume 5 to 10 percent or more of crops such as rice. On our farm in Hillsboro we struggled with rodents, as they would infest the feed bins. We managed to control the problem by keeping a dozen or more cats on our farm at all times. Female cats worked best, as the males were usually too lazy to hunt.

Unfortunately, many farmers apply toxic and environmentally damaging chemicals in attempts to solve their rat problems. Fortunately, agricultural research is starting to show progress in the development of ecologically sound rodent control—by means other than expanding the female feline population. Grant Singleton, a rodent expert at the IRRI in the Philippines, has developed a system in which farmers coordinate as a community to increase hygiene around their rice fields and villages, adjust rice planting dates to disrupt the rodent biological cycle by affecting their food supply at crucial times, and reduce chemical use to encourage natural predators such as birds of prey. More than 100,000 farmers in Vietnam and 75,000 farmers in Indonesia have applied his approach, resulting in reduced losses.

In the prologue, I described the demise of a Holstein, much to the sorrow of Angela and Juan. Cows can be delicate creatures, but they also are highly suited to certain environments. Improved technologies can raise the productivity of cows and help them provide nutritious food and income. For example, with my colleagues Jeff and Darrell, I recently took several Virginia Tech undergraduates to Ecuador to survey farmers about their crop production costs. With logistical support from the Ecuadorian National Agricultural Research Institute (INIAP), we conducted interviews in Bolivar Province in a steeply sloped watershed at about 11,000 feet above sea level. Due to its high elevation, the area is cold and primarily suited for potatoes and pasture. We hiked from farm to farm, often interviewing a farmer in her potato patch or pasture.

I had visited this area several years earlier. It is one of the poorest in the country, but on this trip I was impressed by the progress I saw. Many of the mud and thatch houses had been replaced by brick houses with tin roofs. The cattle were more numerous, looked healthier, and were grazing on more lush pasture than before. INIAP, with support from international donors, had implemented a project to improve potato and pasture production. The resulting gains in the productivity and nutrient content of grasses allowed farmers to raise cattle with improved health and to rotate their fields in a manner that increased potato productivity. In addition, the government provided low-cost loans for houses that the farmers could now afford.

We had broken into small groups for the interviews, and Lauren, Trevor, and I spotted a farmer we wanted to interview. She was tending about five cows and young stock high up in the pasture. We walked up the slope, cautiously eying two barking dogs, and introduced ourselves. By chance, her name was Angela, and she was accompanied by her mother, named Maria. We explained why we were there and inquired if they would be willing to answer our questions. They said yes, and after the interview, the students asked them to explain the dairy production process.

Angela described how she moved the cows around to various locations in the pasture with ropes so they would not overgraze in any one spot. She milked them and then made cheese right in the field. She let me videotape as she described and demonstrated the process. Maria stood by and laughed, perhaps at the thought that anyone would be so ill-informed as to not already know how to make cheese. Unlike Angela, who was dressed in black pants, soccer shirt, rubber boots, and knit hat,

Maria wore the traditional red blouse, long black skirt, and sweater. She wore a red scarf draped over her head to protect it from the sun as well as the strong cold wind that blew constantly on the mountainside.

Angela began the cheese-making process by washing the buckets and milking the cows by hand. A packet of enzymes was placed into the large bucket of milk to separate the curd from the whey. Angela washed her hands and arms and stirred the milk by hand. After a half hour, she stirred the milk again and the cheese solids separated from the liquid whey. She then packed the cheese by hand into long metal tubes in which it would be transported to her house. Most of the milk was processed into cheese due to the difficulty of milk storage and transport. Buyers stopped by to purchase the cheese and take it to the nearby town of Guaranda, although they faced limited competition. After the demonstration, we consumed a large quantity of fresh cheese at the insistence of Angela and her mother.

Livestock research has raised the productivity and well-being of many farmers around the world. Improvements in health, nutrition, management, marketing, and breeding are all important, perhaps in that order. Little can be gained by introducing improved breeds into an area until basic problems with diseases, sanitation, and nutrition are solved. Problems such as foot-and-mouth disease, tsetse-transmitted trypanosomosis (known as sleeping sickness in people), bovine tuberculosis, numerous other diseases, intestinal parasites, and poor pastures limit cattle productivity. Similar types of problems occur with other livestock species. Fortunately, progress has been made through concerted efforts by national and international organizations. The International Livestock Research Institute, headquartered near Nairobi, Kenya, has been particularly helpful in Africa, and national agricultural research systems, universities, and private organizations have tackled a wide range of livestock problems around the world. Research to improve marketing efficiency of products such as milk and cheese has become more important as other livestock problems are gradually being solved.

Agricultural research is a major determinant of growth in agricultural production and protection of previous gains. Better seeds and other inputs raise productivity and reduce pressures on the natural resource base. Without the gains—and losses averted—resulting from agricultural research over the past 50 years, an additional 100 million acres of cropland, about the size of Germany, would have been needed to feed the world.[50] These gains have often saved highly erosive fragile soils and reduced carbon emissions. Those countries with minimal growth in agricultural productivity have generally remained poor, especially countries in sub-Saharan Africa.

New agricultural technologies unleash a diversity of distributional effects on producers and consumers. Consumers usually gain as the extra food reduces prices. The poor, who spend a high proportion of their income on food, especially benefit. Producers may gain due to a lower cost of production, but only if their input costs per ton of output drop more than the output price. Farmers who fail to adopt a new technology may feel the price drop, but not the lower cost. These changes differ by region, farm size, and commodity.

Some people argue against modern agricultural technologies because of the myriad of changes wrought. Often these are people who are already well-fed. That is not to say that all technologies are desirable. Clearly they are not. For example, synthetic pesticides that kill a broad spectrum of pests often cause harmful health and environmental impacts that could be avoided with alternative pest management practices. On the other hand, marker-assisted breeding techniques, which speed up and allow for more precise plant breeding, can have major food security and economic benefits. For example, wheat, the second most important food crop after rice, is continually attacked by new strains of virulent diseases. Without frequent release of new disease-resistant varieties, another six million tons of wheat would be lost annually and millions of more people would go to bed hungry. Fortunately, Cornell University, with support from the Bill and Melinda Gates Foundation and the United Kingdom Department of International Development, has teamed up with the International Maize and Wheat Improvement Center and agricultural scientists around the world to develop improved wheat varieties with durable resistance to wheat rust diseases.[51] Projects such as this one are essential for meeting the food security needs of the poor.

It is important to select carefully and invest wisely in agricultural research. But without continued agricultural productivity growth in developing countries, the increased production required to feed the growing population will not be realized, food prices will rise, and water conflicts and other environmental problems will worsen. Income growth will be lower and, ironically, population growth even faster than it otherwise would be. People live in a changing environment, irrespective of the availability of new technologies. A new environment pressures them to respond and adapt. Their survival depends on it. New technologies can help. Some people in the developed world hold a romantic view of traditional agriculture. Being trapped in hunger is not romantic.

10 It Takes a School

Children are lined up for roll call in front of a two-room school on a rural mountainside in Colombia. Some are clean, some dirty, some tall, some short, some young, some older, and the number of boys roughly equals the number of girls. The teacher is young, but she has finished high school and has some additional training. Most of the kids will complete the fifth grade, a growing number will attend high school in town, and a few will even attend college. This scene is repeated over and over as educational opportunities continue to expand in rural areas of developing countries. It bodes well for their rural development and for slowing population growth, as girls in high school delay marriage and later have fewer children than those without schooling. Education is crucial for agriculture and for overall economic development.

Unfortunately, education still lags in some countries, especially in sub-Saharan Africa. For example, in Niger, Mali, Burkina Faso, and Chad, less than a third of adults over 15 years old are literate, years of schooling averages less than two for adults over 25, and female literacy is less than one-third that of males.[52] But in many remote areas, the situation is gradually improving. Even in places like Afghanistan and Pakistan, mean years of schooling have roughly tripled over the past 30 years.

Unfortunately, years of schooling tell only part of the story, as consistent class attendance of pupils, and even teachers, varies significantly around the world. There is strong evidence that attendance and classroom performance are affected substantially by health problems associated with intestinal parasites, malaria, and other illnesses. Fortunately, something as simple as de-worming pills taken as seldom as three times per year can greatly reduce intestinal parasites and boost attendance,[53] and the bed nets mentioned in chapter 4 are beginning to reduce malaria.

Many schools, even in the poorest regions of the world, ask pupils to wear uniforms and purchase their own school supplies. Students who cannot afford these items are hesitant to attend school. One way for governments and aid agencies to encourage attendance is to help pay for these items. Another, albeit more expensive, solution is to tie cash payments to school attendance, as mentioned in chapter 3. A typical conditional cash-transfer program such as the ones in Mexico and Colombia pay poor mothers if their children maintain at least 80 percent school attendance. Such programs have proved to be quite effective.

The step from primary to secondary education is huge, as families must incur high costs for additional books and uniforms, transportation, and even lodging. Students also lose the income they would earn were they not in school. But more and more students are taking that significant step and attending high school. Considerably smaller percentages of students make the next leap to higher education, especially in Africa, although college opportunities there are growing as well, if still on a small scale.

In some developing countries, attending college is both a mental and a physical challenge. Shortly after the fall of the communist regime in Albania, I visited the Agricultural University of Tirana to help administrators assess how large their university should be and how many students it should allow in each major, given the needs of the new economy. During the visit, I noticed that students there lived in dormitories without mattresses—sleeping on boards that covered the bedsprings—and in rooms without window panes despite temperatures hovering just above freezing. They had blankets, but books and computers were scarce. It takes a dedicated student to perform well under such conditions.

Fortunately, conditions are improving in many colleges and universities in developing countries, even in Africa and South Asia. For example, Tamil Nadu Agricultural University in southern India has teaching and research programs that rival those in more developed countries. Makerere University in Uganda has grown and improved, and its agricultural faculty and students collaborate with universities and faculty in the United States and other countries.

Many universities in Asia and Latin America are now granting high-quality graduate degrees as well, although most students from developing countries who earn PhD degrees still receive them in developed countries. One downside is that many, and for some countries most, of these students take jobs in the countries where they earned an advanced degree and represent a significant resource that is lost to their home countries. However, as economic development continues and incomes rise, increasing numbers of students will return home. Of the more than 40 international students I have supervised for advanced degrees, about half returned to their home countries and the rest were hired by US firms and universities or by international agencies.

One of my students was Pedro Pablo Peña. Pedro grew up in a humble home in the small rural town of Villa Elisa in northwestern Dominican Republic. One of six children, he worked hard from an early age and made friends easily. His father was employed in public works, but most of his relatives and friends were farmers. Pedro rose early in the morning to help his godfather milk cows. He also enjoyed helping his uncle farm, and decided he wanted a career in agriculture that would improve lives in his community. His family could not afford to send Pedro to college, but he studied hard and one day learned that the agricultural college, the Instituto Superior de Agricultura (ISA), near Santiago was offering a scholarship exam in the nearby town of Monte Cristi. He took the test along with hundreds of other students, was selected for an interview, and was one of 38 students awarded scholarships to attend ISA. After graduation, he spent a decade as an agricultural extension agent in the small town of Villa Fundación in southern Dominican Republic. Through his job, he helped bring new technologies to farmers, such as the means for managing their crop pests.

Pedro then began studying agricultural economics and earned a US government funded scholarship to attend Virginia Tech for a master's degree. When he returned home, he headed the training program for a nonprofit organization devoted to agricultural and forestry development. He developed several short-term training programs as well as 10 graduate degree programs at Dominican universities, through which hundreds of students were trained. In 2007, he was named deputy secretary of agriculture by the president of the Dominican Republic, and was responsible for agricultural planning, policies, and data collection, among other tasks. Three years later, he joined the Food and Agricultural Organization of the

United Nations (FAO) as resident representative to the government of El Salvador and then to the government of Ecuador.

From humble beginnings to national and international agricultural development specialist, Pedro accepted help, gave help, and never forgot his roots. Several years ago he helped establish, with a Peace Corps volunteer and with local leaders, a nonprofit rural development organization named APRODEVE in the town of Villa Elisa. That organization has funded and built a community center, town hall, school, first aid clinic, fire station, and cemetery. Returns from education are high and broad.

The value of education can take time to be fully appreciated. While working in South Dakota with the Sisseton Wahpeton Sioux, I noticed that few members of the tribe had college educations. Those who did, such as the planning director, were a tremendous resource for the tribe. One day I paid a visit to the dean of the College of Agriculture at South Dakota State University to discuss the number of Native Americans at SDSU. The university was close to the reservation but had almost no students from the tribe. He said he was aware of that fact and would like to change it, but few Native Americans applied to SDSU. He told me he would guarantee five full scholarships for members of the Sisseton Wahpeton Sioux if I could find five well-qualified students.

I was quite pleased as I drove back to Sisseton that night. It was early fall and Sisseton High School had several Native Americans in the senior class with good grades. I went to the school and spoke to the students. Almost immediately five candidates expressed interest in applying to SDSU for a free college education. Again, I was pleased as I thought this could be the start of something big for the tribe.

My joy was short-lived. One by one the students changed their minds over the next few weeks and decided to either seek employment or join the military. They each told me that their parents asked them not to go. I queried Ed Red Owl, the tribal planning director, about why the parents were hesitant to allow their kids to receive a college education. He said that history had shown that when students from the tribe went away to college, they seldom returned to the reservation, especially if they did well.

Since that experience more than 30 years ago, attitudes about education have changed among many tribal members. A tribal community college now provides degrees aimed at specific jobs on the reservation, but more significantly, a higher education endowment fund was started by the Tribal Council to support students who go to two- or four-year colleges or graduate school. Its aim is to increase the number of tribal members going to college, regardless of field of study.

There is no question that local and family cultures affect attitudes toward education, wherever one lives. I am not saying that every child needs to go to college, but when parents tell their kids from the time they are little that they will go to college if they so desire, even if the parents have no idea of where the funding will come from, it shows they value education. That attitude, combined with scholarships and loans, can change lives when cultures allow. As the creation of the Sisseton higher education endowment fund demonstrates, cultures change over time. They change most rapidly when the value is evident to the local community as well as the individual.

Not all education that can change agriculture involves formal schooling. Adoption of new agricultural technologies in developing countries is crucial for raising incomes and reducing hunger. The private and public sectors each have educational roles to play in making farmers aware of new farming methods as soon as they are available and in helping farmers understand how to use them.

Many groups are involved in this task, but unfortunately, for various reasons, many technology diffusion or "extension" systems around the world are weak. Bangladesh has only one extension agent for every 1,000 or more farmers, so agents are spread thin. In most countries, agents are poorly paid. Some countries have eliminated their publicly supported agricultural extension systems. Nonprofit, nongovernmental organizations fill part of the void, but are usually focused on only a subset of the farmers. Some technologies are spread by the private sector.

Successful extension systems do exist, but few are as effective as that run by the Colombian Coffee Growers Federation. For more than 80 years, it has taken a holistic approach in supporting coffee farmers with agricultural research, technology diffusion, market information and price stabilization efforts, improved roads and schools, electrification, safe and adequate water supplies, health centers, and coffee processing facilities. It monitors and controls the quality of all Colombian coffee exports, and the revenues from selling Colombian coffee go back to growers and their families. Roughly 500,000 growers, or about half of all coffee growers, pay the Coffee Federation a tax on the coffee they sell, and millions more in Colombia's rural coffee communities benefit from the initiatives supported by the Coffee Federation, such as schools, roads, and other social programs. Joining the Coffee Federation is voluntary, but both members and nonmembers receive benefits.

Technology is diffused by hiring and training extension agents raised on coffee farms and assigning them to offices in rural towns throughout the coffee zone. Roughly 1,000 extension agents help the growers learn about new technologies. The agents make farm visits, have radio programs, and most important, hold meetings with groups of coffee growers, contributing to social cohesion. The topics covered in the meetings are jointly planned between farmers and agents and can include any issue associated with their crops and livestock. If the farmers raise concerns, extension agents pass them on to management. There are programs to teach growers how to resolve conflicts, improve their diets, and vaccinate their cattle. Extension agents adapt their efforts to the needs of the communities. The structure of the organization is democratic, with growers having a voice. When an earthquake hits, the federation is there rebuilding farms, mills, and rural infrastructure. When prices drop sharply, the federation uses its savings, to the extent possible, to cushion the impact and provide a safety net for growers.

When Marj and I worked with the Coffee Federation during Peace Corps, we were a tiny part of its effort to address the breadth of farmer needs. Vegetable gardens, fruit trees, rabbits, fish, and chickens met nutritional needs and diversified incomes a little. When Marj brought the veterinarian out to instruct women's groups on how to vaccinate chickens, it reduced the local incidence of a disease called Newcastle. That disease had been sweeping through the countryside every couple of years, wiping out every chick in its path. Vaccination proved a tremendous boon to the economic and nutritional well-being of farm families.

Countries that fail to develop the skills and knowledge of their farmers and their families find it difficult to develop anything else. Utilization of new technologies and other forms of information depends on the workforce being aware of them and understanding how to use them. Better educated individuals earn higher incomes, and these incomes reflect greater productivity. Education helps farmers acquire, understand, and sort out technical and other types of information. It helps people deal with change, and it raises the standard of political debate.

Investment in rural education yields returns not just for farmers but also for society as a whole. When education levels increase in a village, all villagers tend to learn and gain from their more productive neighbors. Rural education helps children of farmers to acquire off-farm jobs, and many children send money back home. Education of girls can be particularly beneficial for development. Educated females live healthier lives, have fewer and healthier children, and earn more than those who do not receive education.

11

Policy Pitfalls

A cow in an apple orchard is like a kid in a candy store. One of our cows would occasionally leap over the fence between the pasture and the apple orchard and gobble as many apples as possible before jumping back into the pasture. The problem is that eating dozens of green apples affects a cow as it does a human: it generates a major gas attack.[54] To make matters worse, the bovine GI tract is more complicated than the human gut because a cow has four stomachs. As a result, food passes through slowly and gas from fermenting apples can build up and cause the cow to bloat. Before the gas can escape, causing her bovine sisters to give her that "OK, who did that" look, she blows up like a balloon.

One day when our cows came into the barn to be milked, Number 9 staggered in, bloated, with her tongue sticking out—not a good sign. She looked as if she would literally explode at any moment. This particular cow had a reputation for jumping the fence, and this time she really paid the price for her athletic prowess. If the pressure in her stomach was not released soon, it would push on her diaphragm enough to prevent her from breathing and she would die. My father called the local veterinarian for help, but he was visiting another farm and could not come right away. Fortunately, my father had seen old Doc Bosshardt attack this problem before. My dad grabbed a hunting knife, felt around on the side of the cow for a small triangular area just above her ribs and stabbed her. He cut through the skin and abdominal muscle into her rumen (stomach). When he pulled the knife out, the gas escaped in an explosive rush that made us gag. The cow shrank back to her normal size and was out of immediate danger. The vet arrived later, cleaned and treated the wound, and Number 9 survived.

Sometimes drastic measures are needed to correct problems created by overindulgence, but those measures must be carefully conceived and administered lest the patient suffer or die. The lesson holds for nations as well as cows. National problems can be economic, and government or international organizations must be equipped with enough knowledge so that they do not worsen the situation with the attempted remedy. My father could have killed our cow if he had stabbed her in the wrong place. Ill-conceived policies can do likewise to economic development.

Left to their own devices, unfettered markets can stimulate self-interested people to act in ways that benefit society as well as themselves. But access to information is unequal, firms can collude, and some people inevitably try to benefit unfairly at the expense of others. Governments are aware of these problems, and may institute policies that attempt to offset the damage, but even well-meaning policies may just make the situation worse. It is not hard to kill the cow. And in many cases, government policies are simply designed to allow leaders to loot or to raise revenues for other public activities.

Farmers respond to price signals, and most governments manipulate prices of farm outputs and inputs through marketing policies, export and import policies, subsidies and taxes, and many other means. They pursue credit policies and land reforms as well. All of these actions affect production, poverty, hunger, and health for better or for worse.

The government of Senegal takes a hands-on approach to the purchasing, storing, processing, and exporting of peanuts. A few years ago it set a common price across the whole country for peanuts that would be processed into oil. Outwardly, it seemed fair but was not. Peanut buyers should be allowed to pay farmers less for peanuts that are produced far from the processing plant than for those produced nearby, in order to cover transport costs incurred by the buyer. With a common price, many buyers simply ignore the distant supply markets, leading farmers to stop producing or to find other means to market their peanuts. The peanut price paid to farmers in Senegal is often set low so the government can profit from peanut exports. The low price causes farmers to reduce peanut production. At times, the government subsidizes inputs such as pesticides and fertilizer to offset the low peanut price, which leads to inefficient use of these products because farmers use more than they should.

Senegal is not alone among developing countries in distorting producer incentives. Sri Lanka currently gives fertilizer away for free to rice farmers, which causes them to plant rice in areas where other crops would be more suitable. Developed countries such as the United States also implement policies that affect farmers in developing countries. For example, the United States subsidizes its cotton producers, which stimulates increased US production. The extra cotton currently lowers the world price of cotton by an estimated 10 percent, which hurts poor cotton farmers in countries such as Mali. Farmers in Mali get hammered twice, once by their own government policies and once by US policies.

It was my first trip to Mali. When I arrived in the airport near Bamako, my ride failed to show. After other passengers had left the airport, I was standing there alone in a dirt parking lot in front of the terminal. In Latin America, Asia, or many countries in Africa, I would be swamped with guys competing to sell me a ride. However, Bamako is remote for a capital city and flights are sporadic. The only other person in the parking lot was a cab driver in an old beat-up Peugeot who offered to drive me into town. My French is weak, but I managed to communicate the name of my obscure hotel and directions to it. The driver was friendly and made a good impression as he attempted to enlighten me about the positive features of his favorite city, especially its food and the friendly people. I told him I teach courses on agricultural economics, and that unleashed a torrent of comments on Malian agriculture, which he felt suffered too often from drought, low prices, seasonal unemployment, and misguided government policies. I missed much of what he was saying, but he seemed to feel cotton policies were a big problem.

Mali is one of the poorest countries on earth, landlocked and mostly dry. It has a northern city called Timbuktu that is so remote it is often used as a metaphor for a distant place. Most farmers are subsistence growers—they are only able to produce what they need to consume, with little surplus to sell. Cotton is one of the few cash crops, and about a fourth of the population depends in part on income from it. A few farmers have begun to grow organic cotton and have found a niche market in Europe for their product. But organic markets have strict rules and certification requirements that are costly to meet and exclude most Malian cotton growers. Some farmers apply pesticides heavily, as cotton is attacked by numerous pests, but heavy application of chemicals is causing cotton pests to become resistant to pesticides.

One Malian farmer I met said that he and members of his village sold cotton to a state-run company that seldom paid on time. Attempts to privatize the company had been thwarted by its debts, and he feared that price risks would only increase with privatization. Eighty percent of cotton farmers in Mali are poor and can ill afford added risk. The farmer also said that privatizing while concurrently organizing farmers into marketing cooperatives might be the better alternative than simply privatizing. He may be correct, but it hasn't happened yet, and the economic situation for millions remains dire. Poverty is one of many factors fomenting unrest in Mali. Rebel and jihadist groups seized the northern half of the country for several months in 2012.

On the other hand, some developing countries have made dramatic changes in their farm policies and reaped the benefits. I once attended a conference in China on the future of agricultural education in that country. Fifteen of us came from other nations to join 45 scientists from Chinese agricultural universities at the meeting. We made presentations, held discussions, and visited a variety of farms as well as tourist sites. A highlight of the trip was an elaborate dinner hosted by the minister of agriculture in the Great Hall of the People.

One topic during dinner was Chinese farm policies. Chinese agriculture has undergone a remarkable transformation, which began in 1979. For the previous 25 years, farming had been organized around collective teams of 20 to 30 households. These teams were required to sell prescribed quantities of output to the government at set prices. Production in excess of the quotas also went to the government. The teams could adjust inputs, but the government set the acreage used for each crop. This rigid system led to stagnant farm output, which dragged down the whole Chinese economy. There were few incentives to excel.

In 1979, the Chinese government introduced the Household Responsibility System, which restored the individual household as the basic unit of farm operation. Each household leased a plot of land from the collective and, after fulfilling an output quota set by the government, could keep any excess or sell it to the government at a price significantly higher than it received for the quota output.

Households could decide the number of acres to plant of each crop. At the same time, the government developed and released improved agricultural technologies. The duration of land leases was gradually extended to several years to encourage capital investment by farmers. The result was a rapid expansion in farm output, with an average growth rate of about 5 percent per year over 30 years. Basically, farmers responded to market incentives in dramatic fashion, and that response in turn, together with policy changes in other sectors of the economy, helped stimulate broad-based income growth and poverty reduction in China.

The minister of agriculture was excited about the economic improvements he saw in his country. But he said he often felt a weight on his shoulders as he made decisions. With 300 million farmers and 1.3 billion consumers, a mistake could have dire consequences, and not just for his career.

Policy changes in China have been as important as the green revolution to the overall food balance in the world. China is not the only communist country that has reaped benefits from at least partly liberalizing its markets. Vietnam also has. Visitors to Vietnam today see markets everywhere, and if it were not for the party members who attend every meeting, one would easily forget that Vietnam is a communist country.

Ironically, when the Soviet Union broke up and communism fell in eastern Europe, abrupt changes in farm policies, especially changes in land policies, hindered countries in that region for several years. Many Russian collectives ended up in the hands of former party officials who had little knowledge of agriculture and few incentives to run the farms efficiently. Production suffered. Albania divided its collectives into tiny uneconomical farms while sorting out legal challenges by families who owned large farms before the communists took over after World War II. When farm size and ownership patterns change abruptly without reasonable rules in place to guide those changes, severe challenges can occur, especially in input and output markets. Disorder can destroy markets as fast as dictators can. When the nations in the former Soviet Union and eastern Europe went cold turkey as they moved to a market economy, they discovered why institutions matter.

Chinese agricultural reforms improved incentives for farmers, and the national government encouraged provincial and county governments to compete with each other in gradually reforming rules to improve farm productivity. However, China's move toward market liberalization and inclusive institutions is far from complete. Many aspects of the economy remain under the command and control of the Communist Party. Property rights are only partly secure, labor mobility is tightly regulated, and some firms are favored by the Communist Party while people have little say in the political process. If maintained, these constraints may reduce future agricultural and economic growth as they stymie innovation and investment, reduce incentives to work, and increase incentives to concentrate wealth rather than generate economic progress.

Some countries have attempted unusual experiments with farm policies. For 25 years, Tanzania structured its farms around communal villages while setting food prices very low. Food production suffered greatly due to a lack of incentives, and everyone, even in the cities, was forced to find a small piece of land to grow part of their own food, as they could not count on any being available in the marketplace. Fortunately, leaders in Tanzania eventually wised up and scrapped the communal villages and low food prices; farmers responded with increased production.

In Zimbabwe, land was unequally distributed to the extreme during its colonial period, creating social unrest and poverty. After independence in 1980, a series of policies was designed to address the land disparity situation between the many small farms and the few large commercial farms. For the most part, those policies had about as much force in law as political opinions tweeted to friends. Land policy reforms were not vigorously implemented, partly due to a lack of resources to assist willing buyers and sellers. In 2000, during a period of political upheaval, commercial farmlands were violently seized by groups affiliated with the ruling party. Farm output fell sharply. Years later, widespread hunger and poverty still prevail in Zimbabwe, although not all due to land policies. Neither the former commercial owners nor the rural masses are intensely farming most of the land, although some poor farmers have benefited from its redistribution.

Many countries have attempted to reform their land tenure systems to increase production incentives and diffuse social unrest. Some have partly succeeded. One success story occurred in the Philippines when a large plantation was broken up and distributed to about 250 workers on the farm. The former owner retained a piece of the farm, developed a facility for processing onions, and organized the new landholders into a farmer cooperative that now produces more than half the onions in the country. Over the years, excellent management of the cooperative has led to significantly improved economic situations for the new landholders and for the former plantation owner.

Credit policies are often as bizarre in developing countries as they are in developed countries. Credit is essential for agricultural growth, but credit markets are rife with possibilities for misinformation, inefficiency, and exploitation. Credit policies can help to regulate these "money markets," but in some cases they lead to misallocation of resources.

During the early stages of economic development, credit sources are mostly informal: friends, relatives, local money lenders, some stores, pawnbrokers. Terms of the loans vary widely, although money lenders and pawnbrokers charge high interest. Informal credit sources are most important for consumer credit. When a farm has a bad year due to a weather problem or other calamity, the family can survive by selling assets such as clothes, pots, or animals, or by obtaining a high-interest loan from the local money lender or other informal credit source. When a farmer needs a loan to purchase production inputs, informal sources are not much help because the lenders worry that a drought or other weather calamity will affect all of their borrowers within the local lending area at the same time, resulting in too many bad loans. Yet formal sources of credit, such as private banks, are scarce in some rural areas. Therefore, the government may step in with a credit program.

When I worked with the Coffee Federation in the Peace Corps, it ran a "supervised credit program" designed to help coffee growers obtain the resources they needed to buy fertilizer and hire workers to transplant and weed while the coffee trees grew. Supervising the credit meant that the loans were provided in installments and that the farmer received instruction from an extension agent on how to use the funds appropriately to improve coffee production. Our office was headed by an agronomist who signed off each time money was dispersed on a loan agreement. The loans were subsidized and carried a low interest rate well below the commercial rate in areas where such loans were available. The farmers loved the program, and the agronomist knew it.

One day a farmer came into the office and asked to have his loan paperwork signed. The agronomist said he would be happy to sign, but he noticed that the farmer listed a couple of pigs as assets. He said to the farmer, "This loan will help you, and you can show appreciation for my signature by giving me a pig." The farmer was not pleased, but he agreed to provide the pig, as he needed the signature.

A week later, another farmer came in to have his loan paperwork signed. The agronomist noted that the farmer listed some copper wire. The agronomist asked for the wire as a gesture of gratitude for his signature. Again, the farmer was not happy, but he agreed.

This type of petty corruption by the agronomist is endemic in government-subsidized credit programs. The signature was worth something because of the subsidy involved, and the agronomist extracted a personal piece of its value.

Unfortunately for the agronomist, farmers tipped off an extension agent who reported what the agronomist had done to higher authorities in the Coffee Federation. They investigated and fired the agronomist.[55] Such remedial action is unfortunately rare in many developing-country organizations, but it is one reason the Coffee Federation has been successful for more than 80 years.

Several years ago I asked a farmer in Peru to identify his major problem. He said he could not borrow enough money from the government credit program. I asked what the interest rate was for a loan. He said it was 50 percent. That seemed steep until I remembered that inflation was 100 percent in Peru at the time, which meant that the price of the products he sold would likely double during the year just due to the drop in currency value, making the loans easy to pay back. The farmer could not borrow enough because the credit was subsidized and the government-run bank was losing money on every transaction. It had to limit the program and was mostly lending to large farms due to lower administrative costs. This program had given large numbers of loans at election time so the government could buy votes, but most of the credit had dried up since the election. The subsidized loans had also scared away a private bank that might have entered the credit market, because it could not compete with an interest rate that was below the rate of inflation.

The following year my graduate student, Eddie, completed his master's thesis to assess the relative importance of interest rates as compared to credit availability for adoption of improved agricultural technologies in two regions of Peru. Availability won hands down. Farmers could afford to pay very high interest and still produce profitably as long as they could obtain credit. Subsidies made little sense since they reduced credit availability.

Credit availability remains a major issue in the developing world, especially for the poor, including poor farmers. In Bangladesh, an innovative bank called the Grameen Bank has made small loans to low-income women since the early 1980s.[56] The purpose of these "microcredit loans" is to help them start small enterprises. Few commercial banks lend to the poor because they lack collateral and are deemed to be a poor risk. The Grameen Bank, however, has had an extraordinarily high repayment rate on its loans, in part because it only lends to individuals if they belong to a local credit group of about 25 borrowers. If anyone in the group fails to make a loan payment, no one in the group can obtain another loan. Peer pressure from group members encourages repayment. The loans are subsidized because administrative costs are high, but the program has helped millions. The cost of the subsidy, however, has constrained expansion of the program because the bank relies on donations to cover it.

Credit institutions in Indonesia, India, Bolivia, and elsewhere have experimented with nonsubsidized group lending. In some cases, the interest charged has been very high, which has reduced repayment rates and encouraged the most creditworthy borrowers to pay off their first loan and get out of the group so they are no longer liable for the debts of others. The success of microcredit programs has also attracted some lenders who charge exorbitant rates for loans to groups and individuals who are poor risks. They hound them for repayment in ways that would make US payday lenders or collection agencies blush.

Group lending has the advantage of reducing the administrative costs of the lender compared to what they would be for individual loans, but they do not readily reduce the risks for certain types of loans. For example, most reputable microcredit lenders do not make loans to cover the expenses of growing crops because a drought, flood, or other weather-related calamity will affect all group members simultaneously, making peer pressure meaningless. Banks are aware of this problem and consequently do not usually make crop loans, even though they may lend for some livestock purchases. Solutions to this problem involve making loans to a wider pool of people and improving insurance mechanisms to address risk. Insurance can be essential in a credit package for crop inputs.

Marj and I were descending a dirt path in Colombia when we suddenly smelled something sweet. It reminded me of the odor when my grandfather boiled down sap to make maple syrup. Following our noses, we finally came to a small level clearing on the side of the mountain that had been carved out of a forest of coffee bushes above and a field of sugarcane stalks below. A large shed stood in the clearing, a mix of smoke and steam wafting from its open sides. Through the steamy haze, we saw three men in the shed, an elderly gent sitting on a stool, a middle-aged man standing in front of five copper kettles lined up in a row, and a teenager driving two horses that walked in a circle as they pulled a large pole attached to their backs with a harness. Each man wore typical *campesino* clothing of light shirt, dark pants, and small felt hat. We entered the shed and noticed the man on the stool feeding freshly cut stalks of sugarcane, one by one, into a press. As the horses pulled the pole around the circle, large gears turned and the press squeezed juice out of the cane into an open tank. The man standing by the kettles wore a heavy brown apron and used a ladle on the end of a long stick to dip the sugar water from the tank into the first kettle. The kettles were fixed in place over a crackling fire, and periodically the man would feed the flames with dried stalks from previously squeezed sugarcane.

The liquid in the kettles boiled steadily, and the man with the ladle moved the thickening and darkening juice from larger to smaller kettles down the line. Next to the last kettle was a large trough without a fire under it. The brown, muddy-looking sugar was removed from the last kettle, spread out in the trough, and cut into blocks after it had partly hardened. Each block, which weighed about a kilogram, was wrapped in banana leaves after it had cooled and then placed in a large sack with several other blocks. We had witnessed the birth of brown sugar. The sacks of brown sugar were destined for the nearest town where they would either be sold in local stores for use in sugar water or hot chocolate or transported to the nearest city. The boiling process reminded me of the one used on our farm to make maple syrup, except we had an evaporator that allowed the ever-thickening sap to flow through a maze of flues until it was thick enough to be drawn off as syrup. It takes 36 to 40 gallons of sap to make 1 gallon of maple syrup. My brothers still make it each spring.

Sugar is produced throughout the developing world, mostly from sugarcane grown in tropical or subtropical areas. Sugar can also be used to produce alcohol, which may help explain the prevalence of *aguardiente* in Pensilvania, and the burgeoning use of sugarcane to produce ethanol for fuel. In many countries, small-scale sugar-making operations like that described previously have been replaced by massive factories that churn out sugar cheaply—so cheaply that US sugar producers find it hard to compete. They have successfully lobbied Congress for import quotas that restrict sugar imports and jack up the price consumers pay for sugar in the United States. This price hike may be good given the obesity epidemic in our country, but it is not good for sugar producers abroad who receive reduced sugar prices and incomes as a result.

Brazil, the largest sugar producer in the world, converts much of its growing sugar output to ethanol. At least it does not use much corn to produce fuel, as we do in the United States. The growing domestic use of corn for US ethanol production leads to reduced corn exports and increased food prices around the world. Without US subsidies to support producers of corn ethanol, fuel would be far less price competitive than food and cattle feed as a use for corn.

The expected rise in energy prices over the long term may mean continued competition between food and fuel uses for crops such as sugarcane, corn, and soybeans. Research to improve the efficiency of replacing sugar and grain used in fuel production with fibrous waste products such as bagasse (the crushed stalks that remain after the sugar juice is squeezed from the sugarcane) or cornstalks holds some promise for the future, but food and fuel are policy-induced competitors today.

The consequences of nonoptimal biofuel and other farm policies are considerable. People are poorer and hungrier than they would be with better policies. Changing those policies requires ideas, careful assessment of what works and what does not, and political will, the last often being the limiting constraint. As with sugar policies, in some cases the problems and the solutions have a link to international trade. Market-distorting policies in the United States, Europe, and Japan affect trade and make it harder for poor farmers in developing countries to compete.

12

Going Global, Buying Local

Cracks worsened in the walls of the eight-story Rana Plaza building on the outskirts of Dhaka, Bangladesh. Inspectors noticed the cracks and called for evacuation. Shops and a bank on the lower floors closed, but the owners of five garment factories on the upper floors ordered employees to keep working. Production was behind schedule on clothing orders placed by foreign companies. On April 24, 2013, the building collapsed, killing 1,127 people in the deadliest garment factory accident in history. The accident occurred just five months after 117 people died in a fire in another clothing factory in Bangladesh. Many garment workers throughout the country face unsafe working conditions every day, and they are only paid about 25 cents an hour.

Under conditions such as these, many outside observers ask whether people would be better off without international trade, especially in clothing or food. Trade conjures up images of children in sweatshops sewing shirts for Walmart while US clothing factories, unable to compete, close up shop and leave dying towns behind. The fear is that trade means exploitation and unemployment. In addition, concerns are expressed that US farm exports destroy agricultural markets in poor countries when cheap food is dumped on them.

Sweatshops do exist in Bangladesh, and in many other parts of the world. Jobs do move abroad due to international competition, exports do destroy some markets, and accidents are common in countries where codes and regulations are ill-defined or not enforced. Every worker has a right to a safe work environment, and changes are certainly needed.[57] Despite these serious concerns over the negative effects of international trade, and recognition of the potential benefits of local production, gains from trade cannot be readily dismissed. In countries such as Bangladesh, international trade has brought about immense social and economic changes that have benefited millions of poor people, especially women. Workers have flocked to factories from villages near and far to escape low incomes, social constraints on women, and drudgery of the farm. For them, the economic and empowerment gains exceed the costs. As development occurs and population growth slows, wages for low-skill garment factory work will continue to rise, as they have in countries such as China and South Korea.

What is the underlying source of the potential economic gains from trade? Such gains can occur because not all nations have equal access to all types of resources, and specialization can lead to efficiency gains. Therefore, a nation can be better off if it focuses many of its efforts on the products that it produces best given its resources and then trades some of those products for others. If each nation focuses in this way, each can be better off, at least potentially. So a country like Colombia, which has an excellent climate for growing coffee, may be better off focusing more resources on coffee than on corn, even though it could produce corn with those same resources. A country like the United States is better off emphasizing corn. Colombia then trades coffee for corn and the United States trades corn for coffee. Colombia would benefit from emphasizing coffee even if it could also produce corn cheaper than the United States. On a relative basis, Colombia can make more from its coffee than its corn, in part because coffee is more difficult to produce in countries like the United States. Consumers in both countries benefit from low-cost imports.

The presence of potential gains from trade does not mean that Colombian farmers, even in the best coffee areas, should devote all their resources to coffee. Coffee prices swing widely, diseases and insects can attack, or too much rain may fall at harvest. Other countries may implement policies that affect market competition for a product irrespective of which country has a "comparative advantage" in producing it.

But generally speaking, Colombian farmers are better off producing some coffee in areas where the climate is suitable. The coffee farmers with whom I worked usually grew sugarcane, plantain, and a little corn as well. They hedged their bets. However, they were fortunate because their resources gave them options. Farmers living higher up the mountain where it was too cold for coffee had to focus on corn or potatoes, and inevitably they were poorer.

The same advantages in specializing and trading are realized in Central America for vegetables, Thailand for rice, Ecuador for bananas, and Bangladesh for clothing. Producers stand to gain from focusing, up to a point, on what they can produce relatively best. They must be careful, however, because any advantage they may have can change over time due to new technologies, rising labor costs, and other factors. For example, labor costs have increased more in Ecuador than in Honduras over the past few years. Production of coffee and vegetables is more labor intensive than bananas. In response, Honduras has increased its production of vegetables and coffee but reduced its banana production, while Ecuador has continued to expand its production of bananas while reducing its production of coffee.

If a country follows its comparative advantage, some sectors of the economy will decline over time while others expand. People will lose jobs and income in one industry while jobs and income are created in others. As long as there is trade, there is the possibility of significant gains, but the gains are inevitably accompanied by losses for some people. If you are one of the losers you will be unhappy, because few countries adequately tax the gainers and funnel those revenues to the losers. No wonder so many people have a love-hate relationship with international trade.

Looking ahead to 2050, the world will likely need 70 to 80 percent more food than it produces today. India will add another 300 million people. The population of Nigeria will exceed that of the United States. Three-fourths of people in India and China currently live on less than $2 per day, but average per capita incomes are rising rapidly. To the extent that these countries are successful in reducing poverty, they will demand significantly more food. East and South Asia, with half the farmland per person as the rest of the world, already fertilize and water their crops at higher rates than other regions. They will be unable to meet their growth in food demand without increased imports of farm products.

As incomes grow, the mix of foods consumed shifts more toward meats, vegetables, and fruits than toward staple foods such as roots and food grains. Some developing countries have a comparative advantage in meats (e.g., Argentina) or vegetables and fruits (e.g., Brazil, Guatemala), and these countries stand to benefit if they can meet the market requirements.

Farmers in Guatemala were angry because their snow peas had suddenly become worthless. The United States, the primary market for the crop, had placed a moratorium on all imports of Guatemalan snow peas because a leaf miner insect was found in a shipment when it arrived in Miami. US inspectors were worried that an exotic pest might harm US farm production. A $35 million industry in Guatemala was destroyed overnight. It was as if hundreds of snow pea farmers were fired with bills to pay and no unemployment checks.

To convey their anger, farmers heaped the worthless pea pods onto the grounds of the American embassy in Guatemala City. The rotting peas caught the attention of the ambassador and he wrote home for help. His letter was passed on to Virginia Tech from Washington, and we dispatched a scientist from our pest management project to Guatemala to identify the bug and devise a control strategy.

We discovered that the leaf miner was already common in the United States, and after developing a plan to reduce the problem, the US market again opened up to Guatemalan snow peas. The cost, however, was a lost crop for an entire year, the loss of farms for some producers, and accelerated migration to the cities.

Trade policies are necessary to protect a country from invasive species and contaminated food. However, many farmers and government officials in developing countries worry that some trade policies of this type are too strict and are there solely to keep their products out. They may be correct in some cases, although the snow pea case was probably not one of them. The snow pea case points to the need for farmers to diversify production to spread risk, and to inspect their products carefully before they leave home ports.

Many farmers in Africa selling vegetables, fruits, and other farm products to Europe found their markets affected when a certification program with a set of specific farm management practices—EUREPGAP—was established in 1997.[58] In order to export their products to Europe, farms had to be certified by an auditor from a commercial certification company. The protocols were developed by retailer, consumer, and environmental organizations and governments in Europe, but in many cases proved difficult for small African growers to meet, causing some export markets to disappear, at least in the short run. Food safety and environmental sustainability is a major concern in most developed countries, and developing countries that wish to export to them are finding they must adjust their farm practices to do so.

Many trade policies affect prices received by producers—and paid by consumers—by restricting imports or subsidizing exports. Shortly after World War II, a temporary agreement called the General Agreement on Tariffs and Trade (GATT) was signed by more than 100 countries in an effort to reduce trade restrictions and settle trade disputes. Over subsequent decades, negotiations by member countries led to reductions in trade barriers for many products. Unfortunately, little progress was made in reducing barriers on agricultural products, much to the dismay of developing countries. Some progress was made on agricultural trade policies in the early 1990s, along with the establishment of a permanent multi-country trade organization, the World Trade Organization (WTO), that replaced the GATT. Because the WTO has paid more attention to agriculture than did the GATT, many developing countries have joined for the first time. The WTO currently has more than 150 member countries.

At the time the Philippines was debating whether or not to join the WTO, I met with the person responsible for agricultural trade policies and negotiations. He had his hands full convincing Philippine farmers that they would benefit from the relaxation of trade restrictions (as required of members), especially restrictions on rice imports. The Philippines did join the WTO in 1995, but farmers there and in many other developing countries still question whether they have gained much. Ironically, while farmers in many countries worry that they will see little in the way of higher prices, officials in some African countries are concerned that food prices will rise too much if trade rules are liberalized. Consumers might pay more, and the poor might suffer.

When a country becomes a member of the WTO, it agrees to follow specific rules that limit the size of its import and export taxes and subsidies. When a country violates the agreement for a specific product by taxing or subsidizing it too much, countries that feel harmed can complain to the WTO. A panel then decides whether the complaint has merit. If the panel decides it does, the offending country is asked to remove the tax or subsidy. If it does not remove it, the harmed country can retaliate against the offender. For example, a few years ago France restricted imports of US soybeans, and the United States filed a complaint. When the import restriction was not removed, the United States threatened to ban all imports of French wine. The French, protective of their national juice, rapidly removed the soybean restriction.

Progress on relaxing trade restrictions, first under the GATT and now under the WTO, has been slow. Perhaps as a result, many small groups of countries have established regional agreements, such as the North American Free Trade Agreement. Regional agreements to reduce trade barriers are easier to establish than broader ones because they involve fewer countries. Several of these agreements have emerged in Africa, where they may be especially helpful to small, landlocked nations. But the benefits and costs of regional agreements are widely debated. Although they reduce barriers among trading partners, they divert trade away from countries outside the group that formed the agreements and may reduce incentives for more general reductions in trade barriers under the WTO.

The "local food" movement mentioned earlier can be beneficial to small, especially part-time, farms that might not otherwise be able to compete with the megafarms. But how does this phenomenon fit in with the notion that international trade is also helpful?

Consumer demand for local foods provides a market for farms that can provide, even on a small scale, fresh high-quality produce (especially vegetables, fruits, and certain livestock products) on a seasonal basis. Quality is measured in freshness, and in some cases through a trust that the crops were grown without synthetic pesticides and that the livestock were raised in relatively humane conditions. If a tomato is shipped to New York from Central America, or even from California or Florida, it cannot be juicy or it will not survive the journey. Therefore, tomatoes are bred to be a bit hard and are picked before they are ripe—and not juicy—if they are to be shipped any distance. Most chickens in the United States are kept in small cages in buildings with thousands of their feathered friends on factory farms. Some consumers are willing to pay extra for the assertion that their chicken wings and eggs have traveled only a short distance from a small farm where the birds were free to roam and roost.

Many local food products are those with inherently high costs of long-distance transport due to their perishability or high water content. Because they are marketed locally in farmers' markets or under direct contract between producer (or small group of producers) and consumer, transportation and storage costs are kept relatively low. In most cases, a higher price can be charged for these foods, given the quality or freshness factor, than would be charged in a larger retail outlet for nonlocal products. Some consumers are willing to pay a premium for the belief—often correct, sometimes mistaken—that less energy is required to bring the product to market due to lower transport costs. They feel they are reducing their carbon footprint by buying locally. As markets for local products grow along with incomes of consumers, they allow small producers to compete with larger ones for certain products, even though those producers fail to achieve efficiencies of mass production.

As countries develop, nonlocal products, including many that are traded internationally, will continue to meet most consumer demand for food. Efficiencies that arise from trade and specialization, and from producing food products where the climate, land, and other resources are most favorable, are too great to be ignored. These efficiencies are especially high for products such as grains, specialty crops such as coffee and tea, and many processed foods. Grains have relatively low transport costs compared to crops like potatoes and tomatoes, which are heavy with water, and many specialty crops have relatively precise climatic requirements. From an economic standpoint, it makes perfect sense for local foods and nonlocal foods to coexist.

13

Cultural Clues

The first year that I was involved in agricultural research in the Philippines, I met with local scientists to plan a set of activities to be completed over the following six months. Although there was more excitement about some tasks than others, everyone agreed that the plan made sense. When I returned to the Philippines six months later, the scientists had completed only some of the activities. The failure to complete all of them was not due to lack of budget or time—the scientists had not even attempted some of the tasks. I asked one of the scientists why and he told me that the scientist in charge did not agree with all the activities we had included in the research plan. "But," I protested, "He told me 'yes' when I asked if the plan was fine. If he did not like something, he should have said 'no' and we could have rethought the activities."

"I know he said 'yes,'" my friend answered, "but you have to learn when 'yes' means 'yes' and when 'yes' means 'no.'"

That cross-cultural lesson is just one of many that I have learned over the years. In several countries, particularly in Southeast Asia, people seem hesitant to say "no" directly when asked a question, perhaps for fear of offending. But that does not necessarily mean they agree. They say "yes" with varying degrees of conviction, and one must learn to decipher the cultural code to avoid misunderstandings. I suppose it is like in the United States when someone says something with which you disagree. You do not want to offend so you say nothing, only to find out later that the other person thought you actually agreed.

Once while driving on a highway in the Philippines, I noticed a sign that said "No Driving on the Shoulder." Traffic is thick near Manila and many drivers try to save time by creating a new lane. On this day, many drivers did not seem to take the warning sign seriously, as they drove their vehicles past us on the right shoulder. A few miles later, we came to a new sign that said, "Absolutely No Driving on the Shoulder!" Transportation officials understood the culture.

Each country has its own cultures, or patterned ways that people think and act, and visitors must respect them and possess at least some understanding of cultural differences to function successfully. Concepts of time, space, ownership, sharing, acceptable interactions between genders, equality and inequality, food preferences, marriage customs, and other societal norms differ enough to create potential for intergroup or interpersonal conflict. Differences embedded in religion are of course particularly sensitive. I am no expert on culture, but here are a few stories that may provide lessons on where to tread lightly and why culture slowly changes with economic development.

Many years ago, I drove from Blacksburg to Charlottesville, Virginia, to watch a football game between longtime rivals, the University of Virginia (UVA) and Virginia Tech. At half time, the announcer on the public address system said that the UVA band was about to play a song that would illustrate the primary difference between the two universities: culture versus agriculture. Playful offense may have been intended, but I took none. The use of *culture* in the word *agriculture* signifies the patterned ways in which humans work with their crops and livestock. Agriculture in a specific location represents a cultural response to the resources, knowledge, and market opportunities available for farm production. Many new technologies developed to help farmers increase food production represent changes in cultural practices—the ways that farmers practice plowing, planting, weeding, harvesting, cultivating, mulching, and other operations to raise food and fiber. But understanding local culture in its broadest sense is also important for agricultural development.

Late blight devastates potato production in cool, humid areas. Scientists at the International Potato Center in Peru work with counterparts in national agricultural research systems to breed new potato varieties with resistance to the disease. They search through the many thousands of types of potatoes that exist in the world to find ones that can withstand the disease without losing yield. When they find one, they cannot just recommend that farmers grow that potato instead of their current one, because farmers and consumers are very picky about their potatoes.

Some like them big and white. Others like them small and red, purple, or yellow. Some potatoes are good for baking, some for boiling, and others for processing. Potatoes come in a surprising array of flavors. Scientists must identify local preferences and then breed late blight resistance into the local favorites.

Some might say that poor people should not be so picky. Food is food for the hungry. That may be true for the extremely poor, although everyone has food favorites. When famine beset South Asia in the 1960s and the green revolution released a high-yielding rice variety called IR-8, many said it would not be planted or eaten in India because its cooking quality was inferior. But people were hungry, and they grew the rice and ate it. The green revolution spread at an accelerated pace, however, after the release of rice varieties with improved cooking quality. Even poor farmers grow food for the market, and the most preferred foods fetch the highest prices.

Many food preferences have developed to maintain human health, including some preferences that have been enshrined in religious customs. It is no coincidence that people in hot climates often prefer spicy foods, and people in cold climates often enjoy relatively bland diets. The capsicum found in hot peppers has antibiotic properties. Bacteria multiply rapidly in food left out in the heat, and spices can help to kill them, thereby reducing the potential for diarrhea. Salt also acts as a preservative, and is often heavily used in places where refrigeration is lacking. Pigs living in parts of the Middle East can spread parasites that are deadly to humans, and they compete with humans for grain. It is probably not a coincidence that two of the predominant religions in the region forbid pork consumption.

In India, many Hindu would rather starve than eat a cow. There are many theories about how the cow became so sacred, but one explanation is that thousands of years ago, farmers began to recognize the many and long-term benefits that living cows could provide. For example, cows available for breeding could produce oxen. An ox was the family tractor and mode of transportation. Those who ate the family cow lost this important power source for their farm. Those who decided not to eat their cows could respond better to natural disasters such as drought because the cows provided products such as milk, as well as fertilizer and cooking fuel from cow manure. Over time, more and more farmers probably avoided beef until it became taboo to eat it. Later the priests stepped in to codify the ban in scripture.

Price often comes into play in food preferences. The southern cone of South America has excellent resources for growing cattle. As a result, beef is cheap and people in countries like Uruguay and Argentina consume it like there is no tomorrow—which there may not be once it hits their arteries. Perhaps a religious taboo will follow.

Resources in Asia are well-suited for growing rice. Some Asians I know have withdrawal symptoms if they go a day without rice. Consumers in certain African countries experience similar symptoms when they go without cassava—most Nigerians need their daily dose of *gari*. Scientists must consider such preferences when conducting agricultural research. Scientists in Nigeria are attempting to reduce adverse health outcomes associated with deficiencies in vitamin A and iron by biofortifying cassava with these nutrients. Similar efforts are underway with rice in the Philippines and India and with sweet potatoes in Kenya.

The second day we were in Pensilvania, Marj and I traveled into the Colombian countryside to meet farmers. We were accompanied by the local extension agent, Luis Carlos, who introduced us and later helped explain to everyone what we had just said in broken Spanish. Marj had packed a lunch. It would be the last time we were so foolish. When we arrived at the first farm, we were given hot chocolate and *arepas*. At the next, they gave us rice, plantain, *arepas*, and hot sugar water with lemon. At the next, they gave us a bottled drink called Pony Malta with a raw egg dropped into it—a most unusual flavor and consistency. At the next stop, we met with the farmer for about an hour, and when it seemed like it was time to go, Luis Carlos said we should wait because they were fixing lunch for us. I quietly told him that we had brought lunch, were already full, and that these folks looked pretty poor. He responded that they would be insulted if we left before lunch. We stayed and they gave us plates of rice crowned with a fried egg, fried plantain, and more *arepas*. The generosity was killing us. Death by *arepas* and rice! It was our fourth meal and it was only 1:00 PM. Marj and I spent the next two years trying to maximize our work time while minimizing our food intake. People were so generous.

It took a while for the sharing custom to rub off on us. At the end of each day, we retreated to our apartment in town to eat our dinner together—Marj, me, and at times Luis Carlos and a few other friends. On the weekends, when many farmers came to town, we met with them in the office or cafe, but seldom invited them to dinner. We accepted their food without sharing our own food or money. We rationalized that we were there to help them with gardens, rabbits, chickens, fish ponds, and nutrition, but not to give any resources other than our advice and labor. We worried it would get messy, especially if we started to give money, because we would not know if farmers sought us out for our technical expertise or our money.

Occasionally, we did feed Luis Carlos, a few other friends, and a couple of kids who were hungry. One kid named Fidel would sometimes follow us into the house when we returned from work. It became a little annoying, but he was hard to turn away. In hindsight, I suppose we were stingy in not opening up our home more than we did, but living in another culture also made us treasure the times when just the two of us were together for dinner.

May 13, 1973

Dear Folks,

Today is Mother's Day here. Everybody is in fiestas. Happy Mother's Day Mom!

Lately, we have begun to feel like we are running a hotel. We told the *Cafeteros* extension agent from San Daniel that he could stay with us when he is in town. He is required by the *Cafeteros* to come to Pensilvania periodically to check in, but they do not pay his food and lodging. He has been here the past three weekends. Yesterday he took off with the veterinarian and we thought we would be alone, but then Luis Carlos showed up. Now he is staying here until the other guy gets back. Also this week we invited a guy who builds *Cafeteros* schools to stay here one night. He also came to dinner and brought a friend. If that were not enough, we now have another kid, besides Fidel, who invites himself in for dinner. He just sort of follows George in the door at night. He came twice this week. We have been cooking for at least three people every night this week and often for breakfast. . . . It would nice to be alone more. . . .

Love,

Marj

Despite having few farmers over to eat, we were sought after anyway to make home visits, and we were always booked two weeks ahead for farm visits. Marj would travel to a rural neighborhood and visit a household each morning and then meet with a group of 20 to 30 women in the afternoon. She was given their most prized meat, chicken, almost every day. I often visited multiple farms per day and never lacked for food.

Marj and I were walking one day in the *campo* on our way to a meeting, and we stopped briefly to check on a garden that I had helped a woman construct and plant the month before. I had told the *señora* we would pass by, but would only stop for a minute to check on the garden. We felt terrible when we arrived and she had prepared a special meal for us, including chicken and a special dessert she had purchased in town. This woman was extremely poor and lived with several children in a one-room shack with a semi-attached kitchen on the outside. We had to eat quickly because we were late for the meeting, and she seemed so disappointed that we could only eat a small part of the feast she had prepared. At least her children ate well that day, but we felt bad having to eat and run.

In my travels since Peace Corps, sharing of food has been a consistent theme, but other types of sharing abound. My graduate student from Nigeria invited Marj and me to dinner. After we ate, she showed us a video of her traditional wedding—or at least the highlights of the 24-hour event. Everyone showed up from her husband's village, and each received a gift based on a request submitted beforehand—items such as soap and toilet paper. Without such sharing, the couple would not have been allowed to marry.

Sharing takes many forms. When jogging one morning in Harare, Zimbabwe, I noticed a street cleaner who was giving coins to a homeless person. When working with the Sioux in South Dakota, I learned that each tribal member with a job was expected to share much of his or her earnings with an extended set of relatives. Agricultural labor-sharing arrangements are common in Africa, as they were on our farm, with uncles, cousins, and neighbors helping on each other's farms. In many rural societies, failure to share would be shockingly antisocial behavior. The group stands or falls together.

Sharing among family members, friends, and even strangers works well when social and economic relationships are highly personal and local. Food, money, and labor sharing are especially prevalent. Sharing reduces risk and facilitates social stability and survival. As countries undergo economic development, social and economic relationships become less and less personal, which necessitates a movement toward increased formal rules for social interaction. Informal rules are great when everyone knows each other in a village and antisocial behavior is punished. If someone takes advantage of others, he or she is ostracized, or worse.

When society becomes less personal and economic transactions are increasingly made from afar, as they inevitably are with economic development, social institutions must adapt. Otherwise, some people will take advantage of others and refuse to share. Taxes must be imposed so that sharing is formalized to meet educational, transportation, security, health, and food needs that were previously addressed by local sharing and other informal rules. If information flowed perfectly so everyone knew what was happening, even in an impersonal society, it might be possible to have fewer formal rules. But without such information, some people will take advantage of their freedom from social constraints and will abuse the system. Formal rules (formal institutions) are both the result and the cause of economic development. Many countries that struggle economically have been unable to develop formal institutions fast enough to facilitate broad-based benefits to society. As a result, a few people dominate everyone else. These countries stagnate or suffer conflict.

Institutions that govern property are especially important for economic development, as they directly affect incentives for economic activity and disincentives for conflict. People seldom share assets as readily as they do food. An inordinately large effort is spent attempting to resolve asset disputes, especially those involving land. Every culture has rules that govern rights to assets. Often those rules have been disrupted by ethnic conflict, colonial intervention, population increases, or economic growth. If assets are very unequally distributed or are redistributed through the actions of an outside power, the seeds of conflict are sown. Unless means are found to rebalance property rights, open warfare may result.

Livestock are a key asset in many rural cultures, and social status can be determined by the number of cows owned. Marriages are negotiated in cattle in some cultures. It is not unusual for conflicts to occur over competition for grazing rights and over land that is suitable for both pasture and crops. I once listened to a farmer from a stationary ethnic group in Tanzania speak with great disdain about the Maasai because they would let their cattle wander across his maize fields and destroy them. The Maasai are traditionally pastoralists and have lived in the area for centuries.

They feel they should have the right to graze their cattle wherever they choose. The Tanzanian and Kenyan governments have struggled for years to resolve this cultural and economic conflict.

Many cultural differences are more subtle than cows versus corn. When Marj and I were invited to a New Year's Eve party at the Bogotá home of the family where we had stayed during Peace Corps training, we were told that festivities would begin at 8:00 PM. We knew that an event scheduled for a particular time might not begin until hours later. Therefore, we knocked on their door around 10:00 PM. We were greeted by the mother of the family, who seemed surprised to see us so early. We sat and chatted with her son for a couple of hours until other guests began to arrive.

It took us almost two years to learn to judge Colombian time. It was good training though, as people in many other cultures are also on Colombian time—for example, my doctor and the cable guy in Blacksburg.

Sometimes when I arrange a meeting in a developing country, I receive the response, "10 is fine, God willing." Essentially they have covered themselves if they arrive late or not at all. It seems God is not so willing when it comes to meetings, as rain, snarled traffic, or headaches often create delays.

Over the years I have spent in developing countries, I have witnessed a gradual change in time perceptions, or at least when their citizens deal with Americans. Either the cultural characteristic has gradually changed toward punctuality, or the people have adjusted their perceptions of time temporarily, knowing how impatient Americans can be. I suspect it is a little of both.

Ben Franklin said, "Time is money." In a sense it is. The increased incomes that accompany economic development imply an increase in the value of people's time; so it would make sense that time perception would gradually change with development. And I suppose some may tire of asking their kids to entertain early arriving American guests.

Time perception is just one of many personal cultural differences that require cross-cultural understanding and flexibility. The expected physical response when greeting someone varies, from kissing both cheeks in Latin America to a simple hand shake, to cultures with a total taboo on touching. Personal distance, hygiene, and cell phone etiquette are other areas of cultural difference. For example, sitting in a meeting of scientists in Bangladesh is like working in a call center. Cell phones ring constantly, and are usually answered. At any point, about five conversations are in progress. In one meeting, I did get a kick out of the "We Wish You a Merry Christmas" ring tone on the phone of a devout Muslim.

Other cultural differences are less charming. Several of us from Virginia Tech were invited to the home of my former Nepalese graduate student during a research project in Nepal. A member of our team, originally from India, asked the student's name. Upon hearing it, he looked a little disappointed. The name of my former student indicated that he was from the barber class, my Indian friend said. Although he went with me to the home, I could tell that he had limited interest in spending the evening with someone who was lower in the ancestral social hierarchy than he—perhaps because he didn't need a haircut.

Virtually all countries have social hierarchies, but some of the rules for social and economic interactions I have encountered in South Asian countries are particularly rigid. These rules can hinder development if the brightest minds are not able to receive education and obtain jobs where they can make their greatest societal contribution. My Nepali friend told me he felt shut out of jobs in Nepal that would fully utilize his talents. He has since moved to the United States, where he teaches at a university. Fortunately, constraints of hierarchy in South Asia are loosening with each new generation. This loosening should facilitate economic development as positions in government and industry are increasingly filled on the basis of merit rather than class.

Few cultural differences are as marked, or sensitive, as the informal rules governing behavior within the household or between the sexes. The roles and rights of women in particular with respect to decision making, education, inheritance, property ownership, dress, interactions with strangers, and farming practices differ across countries and within them. Because women comprise more than half of the agricultural labor in the world, limiting their roles and rights can constrain agricultural development and thus economic development—in addition to being a human rights issue.

When I worked with gardens, rabbits, chickens, and fish ponds in Colombia, a high proportion of my time was spent working with women.

My work with vegetable gardens and chickens was mostly with women. I worked with men to construct rabbit hutches and dig fish ponds, but women and children fed the rabbits and fish. Women also cared for pigs and the occasional cow, whereas men worked in the fields with coffee, plantain, and sugarcane. Because I was in a teaching role, I could work side by side with a woman in building a garden without violating a cultural norm.

Now, when working in the Central Luzon in the Philippines, we encounter no difficulty interacting with both men and women as we design and monitor our experiments in farmers' fields. When we work in Bangladesh, the situation is more complicated. Most of the farmers are Muslim and the accepted interactions between women and male strangers are more limited. In some cases, I would be prohibited from even shaking hands with a woman unless she was to offer her hand first. Women hold positions of great authority in the country, but in many settings they are often expected to be subservient to males.

In Nigeria, I can easily interact with women on farms in the mostly Christian southeast, but the story can be different in parts of the mostly Muslim north. The rights and roles of women differ within farm households across and within countries. Although it is hard to generalize, women in much of the Philippines have greater decision-making authority than do women in Bangladesh. Rights vary significantly within African countries, but women often are restricted from inheriting or owning property.[59] Where they cannot, it can be difficult for women to obtain loans needed to purchase improved farm inputs. In some areas, male extension agents are forbidden from talking to the majority of farmers because they are women. This problem is exacerbated when most of the extension agents are male. In areas where women have limited access to schools, they are at a disadvantage in learning about improved agricultural technologies.

Fortunately, cultures change over time and increased opportunities arise for female education. Increasing numbers of women scientists and extension workers are being trained. Intra-household dynamics are also gradually moving in the direction of empowering women. I recently visited a small farm in the Dominican Republic where the wife and the husband each had their own tomato and pepper fields and their own bank accounts. They shared household expenses. Times they are a-changing.

I recently observed a woman in Nepal who had adopted a set of improved agricultural practices recommended by scientists who work on our IPM CRSP project. When the woman said she was making more money with the new practices than she did before, I asked her what she did with the extra income. She explained that her husband had left her for another woman because she had borne no children. She used part of the money for her everyday needs, but saved some in a retirement account because she would have no one to take care of her in old age. She had a determined look on her face—almost as if she was saying to her ex, "I don't need you, you bum!"

It is perhaps impossible to fully understand a culture besides—or even—one's own. Culture includes thousands of big and little things, real and perceived, logical and illogical, that bind a group of people together. Culture helps to provide cohesion and values that foster the continuity of the group. Cultures continually change at the margin, with alterations happening so slowly that people hardly notice. Positive change occurs at the intersection of cultures, but when change comes too quickly, people often fight back.

Culture also complicates change for those attempting to encourage technological or institutional innovation in a culture other than their own. One of the complicating factors is the multiplicity of cultures that may be found within a single country. As countries urbanize, cultural differences tend to develop between rural and urban areas and between rich and poor.

Much of this book focuses on rural agrarian, low-income areas. Despite poverty, food preferences in these areas are complex, and food can serve many purposes besides nourishment, particularly on special occasions. When I worked with the Sioux Indians they insisted on raising buffalo on their tribal farm for powwows. Saving up for wedding feasts, especially in South Asia, can force families to forego adequate nutrition for months. Such food rituals may be partly designed to reinforce cultural norms that provide longer term security.

I could write a whole book on the cultural differences I have observed—many are irrelevant to the topic at hand, and most I probably misunderstood. My own cultural beliefs have been shaped by my family and rural upbringing in the United States, and by the many cultures I have encountered in schools, universities, and the places I have worked and visited. Fortunately, people in other countries seem to be forgiving of my cultural quirks, or chalk up my misbehavior to a recognition that cultures differ. At lunch time in Bangladesh, I am often asked if I prefer to eat with a fork, which they know is my custom. People there frequently eat with a bare hand, the right hand if they are Islamic. I do not always do as the Romans do when in Rome—I often choose the fork. But I try not to use the left hand, so as not to gross them out, since that hand is considered the one to use when cleaning up in the bathroom.

It is impossible to understand all of the cultural differences in one, let alone several, countries. So my strategy is to try to be sensitive to the cultural differences I pick up, and try to be helpful but not pushy. Appreciation of those who are respectful and who listen seems to be common across many cultures.

14

Avoiding Violence

"If you like, I will show you where the military executed insurgents," our Guatemalan friend said as our van pulled into the fenced-in facility near Sololá, Guatemala, a city overlooking Lake Atitlan in the western highlands. The facility now serves as an agricultural research station and vocational training center, but a few years earlier it was a military outpost during the brutal civil war that ravaged the country for more than three decades. After hostilities ceased, the government decided to dedicate it to activities that would improve the lives of local residents. It donated the facility to a local university, and laboratories have replaced barracks, agricultural classes have replaced military training, and field experiments with corn, tomatoes, and cauliflower now occupy the fields where prisoners were shot. I spent a few days at the research station with other scientists from the IPM CRSP, participating in meetings and reviewing progress with integrated pest management on nearby farms. On an early morning jog inside the perimeter fence of the station, I could not help but reflect on what had transpired there a few years earlier.

Widespread violence in Guatemala lasted 36 years and generated such fear of reprisals it has taken almost two decades to uncover the true extent of the brutality. Many people, especially in rural areas, are afraid to this day to divulge their secret scars from a war that officially ended in 1996. The war was rooted in a colonial history and an economy that facilitated plantation agriculture and its need for cheap labor. Spanish and German immigrants established large plantations of coffee and other export crops, hiring indigenous workers from mountain villages. The workers and their families were often treated poorly and were bound to the plantations through national tax and labor laws that benefited the wealthy. A short-lived land reform effort, aimed at providing small parcels of land to the workers, was derailed by a military coup in 1954. That was followed by a civil war that pitted local insurgents against the military.

At first, insurgents drew moral and logistical support from civilians in towns and villages. Eventually the military began abducting and brutally killing anyone suspected of helping the guerillas. Human rights worker Dan Wilkerson documented how the army entered the indigenous mountain village of Sacuchum one weekend, rounded up residents, then killed numerous villagers, leaving behind 52 widows and 100 orphans.[60] This terror tactic eventually worked; the people began to fear both the insurgents and the Guatemalan army, and they stopped providing food and other supplies to the local fighters. By the time the insurgents and the Guatemalan government signed a peace accord, 300,000 people had been killed, most by the government army. The majority of the deceased were unarmed civilians caught between two forces. Guatemala remains one of the poorest countries in the Western Hemisphere.

Grief and fear are difficult emotions to overcome, and many Guatemalans continue to suffer in silence. Some have devoted their energies to community projects such as building schools, health centers, and roads. Time slowly heals the wounds of war, but only if the source of fear is removed or greatly reduced. Fortunately, the government eventually had the wisdom to turn swords into plowshares, at least in Sololá.

Friends of ours, Jim and Lyn, were Peace Corps volunteers in Liberia several years ago. After they left, that country was devastated by two long civil wars. Shortly after hostilities ceased, they returned to help rebuild. Jim was employed by a university and Lyn by a nonprofit organization. They found their work extremely difficult because during the war, most young boys were forced to become soldiers. The boys not only experienced unspeakable horrors, they also never attended school, and violence became their way of life. A whole generation grew up with combat as their only acquired skill. Families, schools, infrastructure, and most institutions were destroyed. It is taking years for Liberia to recover from such a devastating loss of human and physical capital, and the country remains one of the poorest in the world.

Progress has been slow in rebuilding the country and creating employment. As the years have rolled by, residents and donors alike have found it difficult to remain optimistic. Suffering continues, and the danger of a repeat conflict grows. Honest, hard-working people find it hard to survive. For example, Jim and Lyn knew a man named George who was a driver on a YMCA project. When the project ended, George was laid off. Jim and Lyn did not see him for about five years, until he came to their home, frightfully thin, asking for work for a little cash. He had not found work since the YMCA project, and the house in which he and his family of six had rented a room burned to the ground with everything they owned—a common problem in cities that rely mostly on candles for lighting.

Jim and Lyn found George two weeks' work in their house, which he performed with zeal. They helped him replace his burned driver's permit, retyped his burned resume, and informed him when they heard of jobs for drivers. With high unemployment outside of the subsistence farming sector, it was a long wait.

One day, George was offered a job as a driver. He started the same day he was interviewed and drove for 11 hours until 7:00 AM. For the next three days, he drove two shifts each day, from 3:00 PM to 7:00 AM. It was not unusual that a salary had not been discussed and no paperwork completed. Bosses often test employees in this way before telling them their pay rate, and the offered wage is not negotiable, as so many want jobs. On the fifth day, the general manager, who outranked the man who hired George, showed up with a driver he preferred for the job. The general manager told George he had not approved his hiring and therefore it was not official. Not only was he out of a job, he did not get paid. There was no possibility for redress, as the judicial system is only slowly being developed in Liberia, too slowly for those who need legal protection.

People like George in Liberia and other developing countries must take risks to get a job or otherwise better their lives. But jobs can be few and far between. When people are hungry with little hope for change, they are candidates to be convinced to fight again, and the cycle of violence continues.

Violence and conflict create major challenges for any country, but especially if the nation is attempting to escape poverty. Challenges linger long after the fighting ends. Conflict cripples countries, communities, farms, and families. It destroys lives and assets, with numerous short- and long-run effects. Countries are destabilized, creating fertile places for extremists and drug traffickers. Refugees spill over into neighboring countries and worsen health problems. Finding ways to resolve conflict and provide security for populations at risk is essential for alleviating poverty, hunger, and disease. Hopelessness, fear, anger, religious and ethnic differences, cultural clashes, extreme inequality, drugs, and political power struggles under extractive institutions are among the many causes of violence in developing countries. Where nothing is certain, life is cheap, and a culture of violence can persist for years.

Nigeria experienced a horrible civil war in the late 1960s, and violence has lingered since. During the war, economic and religious tensions culminated in an unsuccessful attempt by the southeastern region of the country to achieve independence. The war was partly rooted in a colonial history in which Nigeria was created by drawing a border around nearly 300 ethnic and religious groups. More than 3 million people died in the civil war, many from illness or starvation. The war ended more than 40 years ago, but like a bruise that won't heal, violence has died down but will not die out, especially in areas shared by Muslims and Christians or districts rich in oil. Oil wealth lines the pockets of a small minority of the population in the south, creating resentment in the north, which has seen increased attacks by Islamic extremists. Religious conflict and economic disparity together lead to the periodic slaughter of men, women, and children. In 2012, more than 1,000 people died in sectarian violence. Nigeria spends a quarter of its federal budget on security, yet violence frequently flares up. Unless economic opportunity expands, those flare-ups may eventually lead to explosions.

To Americans, conflict in a nation such as Nigeria, Liberia, the Democratic Republic of the Congo, Burundi, Angola, Mozambique, Somalia, or Sudan seems a world away. We may see it in the news, but soon it is out of sight and out of mind. The country is across the ocean, and the people there are not like us. But of course they are, especially the children. The kids may be ragged and use a field or a street for a toilet, but if not conscripted as child soldiers, they want to play, go to school, and eventually work. Yet for a country in conflict, violence robs the children of their childhood and their dreams—and the nation of its hope for development.

It is simply called "*La Violencia,*" the period of extraordinary violence in Colombia from 1948 to 1958. More recent drug violence and guerilla insurgency in Colombia pale in comparison to the atrocities of that period. In 1971, everyone in Pensilvania remembered. Some would talk about it. Others would not, or could not. According to political scientist Robert Dix, the period of *La Violencia* began with a political assassination in Bogotá, but soon spread to other urban and rural areas, pitting liberals against conservatives with each committing unspeakable acts of violence against their neighbors.[61] According to Dix, the Colombian press fanned the flames with sensationalism and partisan blame for specific incidents. The country entered a downward spiral of competitive cruelty, and by the time *La Violencia* was over, 200,000 people had perished. One farmer told me of a relative whose unborn child was cut from her womb and replaced with a chicken. A friend of another farmer had his throat slit with a *machete*, his tongue pulled through the hole in his neck and laid on his chest like a necktie.

Eventually the warring factions reached an agreement that would give the country 16 years of relative peace as the political parties alternated power every four years. But some warned us to steer clear of Manzanares, a town two hours to the south of Pensilvania. They said violence lurked there just below the surface. Given the friendliness of everyone in Pensilvania, we found that hard to believe, but we stayed away nonetheless.

It is hard to say with certainty why some countries—and towns—are more violent than others. Colombians, from the poorest to the richest—well, I haven't met the richest—are the nicest people I have met anywhere in the world. Yet centuries of inequality and lack of economic progress for some segments of its diverse society have spawned guerilla movements and pitted left against right for decades.[62]

Certainly the drug trade doesn't help, as it partly funds the insurgents and the right wing paramilitaries formed in opposition, but the inability to create an inclusive institutional framework that closes glaring asset and income gaps appears to be a fundamental cause of the violence. Inequality resulting from extractive institutions causes infighting and war. This same phenomenon created civil wars in Central America in Nicaragua in the 1970s, El Salvador in the 1980s, and Guatemala from the 1960s to the 1990s. It has offered fertile ground for the Shining Path in Peru, the Maoists in Nepal, and many other guerilla movements around the world. Economic motives triggering countrywide violence are often mixed with political or religious aims, and in some cases intertwine with reactions to a changing society. The mixture is particularly lethal when the violence is financed by trade in drugs, oil, diamonds, and other such wealth.

I have visited Colombia a few times since our Peace Corps days. One Sunday I traveled there to interview scientists at the International Center for Tropical Agriculture (CIAT) near Cali about their efforts to reduce pests on cassava in Africa and Latin America.[63] My mother called me before I left, lamenting my going to a "dangerous place," as she so often had over the years. Later the next day, April 16, 2007, I was in a meeting with scientists when someone handed me a news bulletin: "32 students and faculty shot and killed at Virginia Tech, many others wounded." Violence had found its way to Blacksburg.

I returned home and for the next several days met nonstop with students—emotions raw. Students created a memorial to the victims in the center of campus, and messages of support poured in from around the country and world. When classes began again, I debated what to say. I was co-teaching a course to prepare a group of undergraduate students to go to Ecuador on a research project. Julia, a graduate student in biological systems engineering, had been assisting on the project and had presented her research to the group two weeks earlier. She was excited and passionate about her research and was planning to return to Ecuador in May to collect more data. She hoped to spend her career solving water problems that constrain sustainable agriculture in many countries. On the morning of April 16, she was killed in hydrology class in Norris Hall.

The tragedy at Virginia Tech taught me how the effects of even a single brief event can traumatize for years. The violence at Virginia Tech was tiny compared to the slaughter in countries such as Syria and Sudan, but the healing has taken time—not only for the families most affected, but for the broader community as well.

The students in my class went to Ecuador that summer, more determined than ever to complete their research, but some of them struggled. Many in the Virginia Tech community suffered from post-traumatic stress syndrome in the aftermath of the tragedy. In the weeks after April 16, Marj showed no overt signs of distress, but her feet soon became so numb she would never drive again. I awoke with nightmares for months, and my productivity suffered. While much healing has now occurred at Virginia Tech, the aftermath of the event brought home to me the hidden toll that violence takes on people who are close to it, even those not physically harmed and only on its periphery.

The mental toll and lost productivity for a country besieged by a major conflict, such as those experienced in Liberia, Nigeria, Colombia,

Guatemala, Syria, or Sudan, are simply immense. The biggest problem associated with violence is not always the immediate deaths—unless you are a victim or a relative of one. It is the damage inflicted by the mental distress, chronic suffering, and long-term productivity loss caused by the violence that can doom whole societies to prolonged poverty, illness, and malnutrition. Time eventually helps heal the hurt, but its effects can linger for years.

The occurrence of society-wide violence so detrimental to agricultural and economic development is often related to the manner in which power is held.[64] In societies in which wealth and power are concentrated in the same few armed people, peace is often fragile and periodically punctuated by violent reactions to injustice. Those reactions in turn lead to violent and costly reprisals. Societies with more open access to power are often more stable. Open, centralized societies with long-standing permanent organizations, rewards linked to performance, and militaries separated from politics are more adaptable to external and internal shocks and less inclined to seek violent solutions. Such societies usually have checks and balances on state power. Poor, slow-growing countries often lack such restraints, and some people exploit the situation.

The United States had its own Civil War, of course. Some argue that the Civil War was inevitable despite carefully designed democratic processes put in place when the nation was founded. The issues of slavery and the relative strengths of federal versus states' rights had to be resolved. So perhaps it is not surprising that developing nations with widespread injustices and fuzzy rules are prone to violence while they refine their institutions. The key for them is to avoid being trapped in conflict with repeated bouts of violence in which new groups replace old groups with little improvement for the masses.

If a nation can escape from conflict, there is some evidence from countries such as Angola and Mozambique that the shake-up in institutions can have a positive long-run benefit, partly offsetting the negative effects of the upheaval. Property rights may become more secure, and trust in public institutions may eventually improve. People exposed to violence may even become more altruistic in their behavior to others than they were before the strife.[65] Simple acts of human kindness can be a powerful force for promoting healing and peace. However, violence can also create a significant drag on agricultural and economic development for years after a conflict has ceased. Business investment and improvements in education, employment, and basic trust can lag behind the cessation of hostilities by several years. People must be convinced that the new institutions provide a dependable structure for their businesses and everyday lives. Foreign assistance may play a useful role in the long process of rebuilding roads, schools, and institutions once the hostilities cease. It has recently supported post-conflict agricultural development in countries such as Mozambique and Nepal, providing improved agricultural technologies to vulnerable farm households and connecting farmers to markets.

15

Doing Well
by Giving Goods

Giving food sounds simple. An earthquake hits Haiti. A drought devastates Ethiopia and Somalia. People are hungry. We send food and water or they die. The benefits of food aid are easiest to see during or directly after a disaster. The need is evident and if food is not sent, prices of available food will soar and the poor may starve. Logistics are complex. Some food aid will be wasted, but so what?—many people will eat and live. Pictures of children starving during a disaster tell a terrible story as refugee camps swell to the size of Seattle or St. Louis. Food aid cannot save everyone; thousands may perish, first the infants and elderly, later the school kids, and finally the rest. In just three months in 2011, an estimated 29,000 children under the age of five starved to death during a drought in the Horn of Africa. But as the world responds to reports of starvation, thousands more are saved with rice, rehydration, and nutritious supplements. Foods like Plumpy'nut, a simple paste made of peanuts and powdered milk, can bring severely malnourished children back from the brink of death to an almost normal weight in a few months. How could we not help?

But assessing the desirability of food and other forms of foreign aid to regions with chronic hunger and poverty is more complicated. If food is given free or at reduced prices, the added food may drive prices down so that farmers lose incentive to produce and sell their food. A country may become dependent on aid and divert resources from producing food to buying guns. Food or other aid may not reach the poor; it may be used for partisan political purposes or for personal gain by corrupt officials. Should individuals be asked to work to receive the aid? The benefits of long-term foreign aid are often far from clear.

Foreign aid is a subsidized transfer of resources—money, food, and services—from one country to another. Most of the resources are provided by governments or international agencies, but many come from nonprofit organizations such as the Rockefeller or Bill and Melinda Gates Foundations. When I ask my undergraduate class what percentage of the US federal budget they think is devoted to foreign development assistance (nonmilitary foreign aid), they typically say 3 to 15 percent. I ask them if it should be higher or lower. Some say lower—about 2 percent. Some are surprised when I then tell them it is currently less than 1 percent. Cutting development aid is not going to balance the federal budget. Still, why should we spend anything on aid? Is 1 percent too much or too little? Does the aid do any good or does it actually do harm? How does it reach those who need it?

Let's start with food. A few years ago, colleagues and I at Virginia Tech undertook a study in Kenya to assess the pros and cons of a long-term food aid program called Food for Work (FFW).[66] In that program, maize and cooking oil were given to people who helped build local roads and farm fences. In our study, we took a random survey of hundreds of people in the region where the food aid was distributed. We compared the incomes and nutrition of food recipients with those of nonrecipients and came to several conclusions.

First, only the very poor signed up for the program as the small amount of food given in exchange for work was comparable to a low wage. Second, the FFW program exerted little downward pressure on the local maize price because the participants did not buy much food before the program began. They were eating the little they could grow themselves. Although they sold some of the food aid they received, the amount was not enough to depress prices. Third, many of the people who participated in the program were farmers, but their FFW activity was mostly during times when little labor was needed on their farms. Fourth, many of those who participated in the first year no longer did so in subsequent years. They used money from the part of the food aid they sold to buy additional farm inputs and then increased the time they devoted to farming, producing additional food of their own. They had worked themselves off the aid, which was one of the objectives of the program. Fifth, the nutritional levels of program participants improved.

The above example illustrates that food aid can indeed have positive effects on families if it is carefully administered, even in a noncrisis situation. However, positive effects are not guaranteed, and careful targeting of food aid implies significant costs as well. Without careful administration, the food can bypass the poorest of the poor and be used to prop up suspect governments, including corrupt local governments. To make matters worse, food aid provided by the United States has strings attached that are aimed at bolstering US farmers and shippers—the food must be bought from US farmers even if cheaper food is available closer to the recipient nation, and most of the food aid must be carried by American-owned ships. These strings make the aid more expensive than it need be. Giving food is easy. Doing it efficiently and in a meaningful way that contributes to long-term development is complicated and requires careful evaluation of food aid programs.

Most foreign aid is not food. Such aid has also built schools, agricultural research stations, and roads; it has improved minds, health, nutrition, and markets. Many agricultural aid projects have provided significant economic benefits. For example, I recently visited several groups of small-holder farmers in Nepal who, with the help of foreign aid, have formed marketing committees and collection centers for their vegetables and other farm products.[67] These committees and centers coordinate production across farms to maximize prices received, bring technical assistance to farmers, improve access to inputs and credit, and provide locations to aggregate produce. Farmers were pleased to show me their marketing plans, micro-irrigation systems, healthy looking plants, and collection center. Inexpensive drip irrigation systems allow the farmers to produce off-season vegetables when prices are high. More than 80,000 disadvantaged households have benefited from this roughly $10 million aid program, which has increased their annual incomes by more than $200 per acre.

Unfortunately, aid is often abused as well. Many years ago, I assisted the chairman of the council that oversees agricultural research in Bangladesh, working with his staff to complete a study of agricultural research priorities. The chairman was a bright and competent leader who was well-respected throughout the research system. One day I arrived at the council building and found that he had "been retired"—at age 45.

It is not unusual for political appointees to be removed on short notice, particularly when they can be made scapegoats for failed policies, but what stood out in this case was the reason the chairman was relieved of his duties. A US-based nonprofit organization with an office in his building had held a two-day conference on the topic of flood control. It seems the World Bank and the French had proposed a massive multibillion dollar project that would basically build concrete walls along the rivers to hold them back during the rainy season.

Flood walls may sound like a good idea—if ecology is not your forte. Even though it rains by the bucket in most years and floods are inevitable, all that soggy weather brings out the best in many plants and animals. Bangladesh is famous for its rice and fish, neither of which does well out of water. And the floods deposit fertile silt, stolen upstream from Nepal and India, on the fields. Concrete barriers would hurt agriculture more than help it. Participants in the conference came to this same conclusion and unabashedly said so in a paper that listed the Bangladesh Agricultural Research Council as a cosponsor of the conference.

Here was the rub. If there is one thing that high-level Bangladeshi officials hate more than wet feet in a flood, it is forgoing a slice of the spoils from a $13 billion concrete construction contract. Politicians in power could not tolerate any evidence that concrete flood walls might not be a good idea. They had too much riding on it. The council chair was canned for even allowing discussion of the pros and cons of the proposal in his building.

It is easy to make a big deal of corruption. It occurs in every country, including our own, and it is just a part of doing business in much of the world. But the concern is the size of the problem in developing countries that have weak institutional controls. Donors may shun those countries, but some of them, such as Afghanistan, are simply deemed too important geopolitically to ignore.

Foreign assistance is provided for a variety of political and economic reasons, as well as compassion. A look at the major countries receiving development aid from the United States indicates that political reasons—a desire to buy friends to improve national security—certainly drive many aid flows. Aid is used to help stabilize fragile nations. And, as developing countries become more prosperous, developed countries benefit economically from these emerging markets. But clearly a concern for those in need is also important, and public officials responsible for implementing aid programs try hard to maximize benefits to local populations.

For almost three decades, aid budgets for agricultural programs declined in the United States, from about 25 percent of all aid in the early 1980s to around 2 percent in 2007. With the shock of higher food prices that began in 2008, the primary agency responsible for US foreign aid, USAID, received an increased appropriation to support agricultural research due to concern over destabilizing effects of high food prices in poor countries. Officials at USAID asked several people from outside the agency to meet with them to discuss the process for allocating those funds under a program called Feed the Future. Their primary goals were to improve food security, health, and nutrition of the most vulnerable populations. In discussions and subsequent actions that exhibited a focus on high-yield research activities in the neediest yet politically stable countries, it was clear the officials were serious about identifying the programs that would provide the greatest chance of achieving the goals. Their actions were encouraging because overwhelming evidence supports the conclusion that agricultural research and education are two of the best investments that developing countries and their supporters can make to advance economic development.[68]

Some of the benefits of foreign aid are derived from technical assistance provided by outside experts, and the benefits flow both ways. Technical assistance at an industry or country level can help stimulate economic growth, and also brings ideas and products back to the donor. Many wheat varieties currently planted in the United States derive their genetic heritage from the green revolution that helped stave off hunger in Asia and Latin America years ago—a win-win situation.

Another example of aid providing benefits for both the donor and recipient is found in a vegetable cooperative in Guatemala named Cuatro Pinos, which was organized in the late 1970s with assistance from Switzerland and the United States. It expanded in the 1980s, but struggled in the 1990s under poor management. In the late 1990s, a USAID-funded project helped it to establish an improved protocol for inspecting produce locally to reduce its products being rejected at the US border. The International Center for Tropical Agriculture helped link the cooperative to Costco, a major food company in the United States that contracts with farmers, most of them very small producers that supply up to 20 different products. The Cuatro Pinos cooperative and Costco work with more than

5,000 vegetable producers, and Guatemalan poverty has been reduced. The cooperative directly employs more than 1,000 workers, and another 12,000 workers are employed on the farms of the local producers. US consumers currently benefit from low-priced produce in the winter as a result of this earlier foreign aid investment.

Another example of technical assistance at the grassroots level that generates benefits in both directions is the Peace Corps. Peace Corps volunteers provide assistance to families and children on farms, in schools, and at health centers, but they also learn about cultural and economic differences and they gain an appreciation for their own good fortune. Being eaten by mosquitoes in a malaria-infested area, taking a cold shower with a bucket, and watching how hard people work for only a tiny amount of income provides a new perspective on the world. When volunteers return from the Peace Corps, they enter various fields in which they educate, legislate, advise or lead businesses, communities, and churches, as well as otherwise influence global perspectives of those around them.

I once met with the minister of agriculture in Peru to discuss an agricultural research program there. Someone told him that a few years earlier I had been a Peace Corps volunteer in Colombia. He said he liked to discuss farm policies and technologies with former agricultural volunteers because they usually had a feel for the way farmers think. He didn't mention whether he liked to meet with gringo agricultural economics professors, but I suspect it is lower on his list.

Peace Corps volunteers in Peru once took a young boy named Alejandro under their wing, supported him through high school, and helped him get to the University of San Francisco, where he earned a bachelor's degree. Alejandro eventually earned a PhD at Stanford. Peace Corps left Peru in 1975, but one of the first things Dr. Alejandro Toledo did upon becoming president of Peru was to call the president of the United States and invite Peace Corps to return.[69]

Foreign assistance to agriculture has contributed immensely to human resource development in low-income countries. Aid has provided long- and short-term education to thousands of graduate students, scientists, and extension workers, and technical instruction to millions of farmers. The result has been increased agricultural production and incomes. Hundreds of schools, universities, and research institutions have been developed in low-income countries, with some universities strengthened to the point that they now provide graduate degrees of their own. I recently coadvised a PhD student, Rukmanika (not her real name), who completed her doctorate at an agricultural university in southern India. A couple of generations ago, it was rare for a woman growing up on a small farm in rural India to have the opportunity to go to school, much less to high school and college. The parents of Rukmanika have a small farm with one cow in the state of Tamil Nadu. Neither well-to-do nor poor, they are school teachers as well as farmers. Rukmanika has an older sister and an unmarried aunt who live with them. The aunt cares for the cow. In a sense, their small farm has transitioned just as the Norton farm did. Agriculture is still a part of their family life, but only part-time now. Most of their income is earned off the farm, and their standard of living has risen substantially.

In the past, the marriage of Rukmanika would have been arranged by her parents while she was still a teenager.[70] She would have quickly borne several children. Today, she is still single and in her mid-twenties. However, not everything has changed. Rukmanika is waiting for her sister to marry first before she will be allowed to marry, and her marriage will still be arranged by her parents. Her education, however, is seen as a plus that will allow her to marry "better" than she might have before, and she will be allowed to work outside the home. Her survival and that of her children will no longer depend on the vagaries of drought, floods, and pests. Unfortunately, many farm families in India are still subject to those vagaries and suffer from chronic malnutrition. Fortunately, the number of malnourished people has gradually declined as the previous generation of Indian scientists, and a new generation exemplified by Rukmanika, have discovered new technologies and institutions to help Indian agriculture improve.

Foreign aid provides food and economic benefits, but it also supplies emotional support, hope, and healing after disasters or violent events. It can be healing simply to know that someone cares. Many who experience the after-effects of a natural disaster or other tragedy come to appreciate the therapeutic effects of others reaching out to them as they try to rebuild their lives after tornadoes, floods, or other unfortunate events.

We came to appreciate the power of such support in the weeks and months after the Virginia Tech tragedy in April 2007. The families of the victims and our campus community experienced numerous acts of kindness from around the nation and world. That kindness helped us cope with the emotional pain. Support came from unexpected sources, including donations from abroad to the fund for victims' families. The New York Yankees contributed a million dollars, and in March 2008 they came to Blacksburg to lend moral support to the community by playing a ball game with the Virginia Tech Hokies.[71]

From my office window I can see the memorial to the victims of the tragedy. On the day of the Yankee game, when a crowd began to gather there, I knew the team was about to arrive. My colleague Jeff and I rushed over, just as their buses pulled up. The Yankees walked around the 32 engraved stones and stood silently as they heard how students had created a memorial of stones and flowers, right after the shootings. The site had become a focal point for remembering and healing.

A student near me said she was a big Yankee fan and that one of the victims had been her fiancé. As the Yankees filed around the semicircle in front of the stones, she asked one of the star players to sign her shirt, which bore a picture of the man she missed. As he signed, I heard her say it was one of the greatest moments of her life. That player later told a news reporter he was never prouder to be a Yankee.

The Yankees headed over to the ball field to warm up—and students swarmed them. After batting practice and a brief ceremony, 32 balloons were released to the sky. The game was played, and Virginia Tech was no match for Derek Jeter and his Bronx Bombers. All the Hokie players got into the game. After each inning, they high-fived and smiled from ear to ear. The score said they lost, but there was joy in Hokieville that day.

When we were kids, listening to Yankee games on the radio made bearable the dirty job of hand picking stones off our farm fields. When the Yankees came to Virginia Tech, their support helped us bear a much greater pain. As they reached out, it raised our spirits and increased my understanding of the emotional benefits of foreign aid.

Victims of chronic poverty suffer enduring tragedy and pain, and foreign aid may help them to feel that someone cares. USAID recently funded an agricultural project with a site in eastern Uganda. The site is located in an area where many people suffer significantly during the hunger season from April to August as stocks from the previous maize crop dwindle, and they wait for their current crop to grow and mature. I visited there in June 2013 with two graduate students and representatives from a local partner organization, Appropriate Technology (AT). We interviewed farmers to obtain their views on new agricultural practices being demonstrated by AT and adopted by some of the farmers.

Following one interview, a farmer took my right hand in both of his, looked me in the eye, and said softly, "Thank you." I asked him why, as it could not have been much fun answering our survey questions for an hour. He said, "Thank you for the new crop practices that help feed my family."

I said, "Don't thank me. You, who took the risk, and those who demonstrated the practices in your village deserve the credit."

He replied, "I know, but I thank you for *the idea.*"

He knew that someone on a foreign aid project had helped design the new practices, and he had been waiting for that person to come so he could say thank you. I simply said "You're welcome." I had not personally designed the practices, but at that moment, it didn't matter. I was the face of foreign aid; he was grateful for the assistance, and he wanted me to know. So I pass along his thanks to the supporters of a program that reduced hunger and raised the spirits of a family in eastern Uganda.

Reaching out to others not only benefits recipients of the assistance, but the donors as well, and not just economically or politically. Individuals and nations can gain emotional satisfaction from compassion. When we send water, money, medical assistance, and other aid to a country like Haiti after an earthquake, it helps us as well as them. An example close to home of the power of helping others is provided by a couple who lived two doors down from us in Blacksburg. Bob and Emily spent their whole lives assisting others. They fought racial and social injustice at every turn. They were always aware of what was happening in the lives of their neighbors and friends, and whenever they saw a need they acted on it. Even as the years advanced and their health declined, they continued to assist others.

Bob and Emily had a basement apartment in their home that they lent out to numerous students over the years, many of them from other countries. Once, a person named Theo, from Benin in West Africa, came to visit Virginia Tech for two months to work on a research project funded by the USAID-funded IPM CRSP. His English was rather poor. Bob and

Emily offered to have Theo stay with them. He did and by the end of two months, his English had improved so much that he was admitted into the graduate program with support from the project. After Theo finished his master's degree, he returned to West Africa to work in agricultural development, appreciative of all the assistance that Bob and Emily provided.

After Emily turned 90, her health declined rapidly. The last few weeks of her life she was confined to a bed, but she remained cheerful and received a long stream of visitors. Marj and I stopped by just before I left on a trip to Ecuador. Emily said she was concerned that Marj would be alone while I was gone. We assured her that Marj would be fine.

A few days later, Marj received a call in her office from Bob. He and Emily had heard that a thunderstorm was predicted at the time Marj would be riding her three-wheel scooter home from work. If it rained, he would come with Emily's caregiver to retrieve her. Marj said it was not necessary; the caregiver was there to help Emily, not her friends. But when a storm arrived in the late afternoon, they came to Marj's office and brought her home. When Marj told the caregiver she felt guilty taking her away from Emily, the caregiver reassured her that her job was to do whatever pleased Emily and Bob, and it pleased them the most to help others. In typical fashion, Emily was on her deathbed, but she was concerned that Marj might get wet. Emily passed peacefully a few days later.

A memorial service was held and, as expected, the church was packed. I had e-mailed Theo in West Africa to inform him of Emily's passing. He sent his condolences and regrets that he could not return to the United States for the service. He said that his family and friends knew all about Emily and that they also would hold a service for her in Benin. So as testimonials were being said in gratitude for the life of Emily in Blacksburg, her life was being celebrated as well in a village in West Africa. Emily and Bob were neighbors to the world.

Reaching out to others with foreign aid or individual charitable actions can significantly help donors and recipients, but if it is administered naively, the opposite can occur. There are many examples of charitable intentions making a situation worse. Good intentions are obviously not enough. But how do we predict if our foreign aid will do more harm than good? First, there is ample evidence that aid given to countries with highly distorted economic incentives tends to benefit the few and not the many.[72] Aid given to countries with highly corrupt and overly autocratic governments has little chance to succeed and may only serve to reinforce policies that hurt the poor. For example, aid has little chance to succeed in a country like Chad where corruption siphons off money intended for public programs such as schools and health care before the money can accomplish its purpose.[73] However, some types of agricultural development aid have a better chance to succeed than other types, and short-term food aid following natural disasters must be attempted for humanitarian reasons regardless of local governance.

After 40 years of observing the effects of aid in recipient countries, I conclude that it has relieved much suffering in short-term crises, and economic improvements have followed when donors have recognized local constraints and used aid to stimulate appropriate incentives for change. Assistance for agricultural research alone has indisputably saved millions from starvation and had a high return on public investment.[74] Aid for roads, schools, and health programs has improved the lives of millions more.

But aid must be transferred efficiently and target public investments that spur rather than stifle local private investment. Aid can be used to stimulate rural nonfarm employment and enable small farms to operate part-time as household incomes rise. However, wherever aid simply props up corrupt officials and lines their pockets, or is insufficiently programmed and just creates dependency, its chances of success are nil. Understanding the behavior of its recipients, with their needs and constraints, is one key to making foreign aid truly useful.

16

Finding Hope

The farming transformation continues around the world. Farmers feed more people than ever before, and those people increasingly have means to consume more than food. Some farms grow and specialize while others convert to part-time status. Agricultural products move farther and faster each year, bringing benefits as well as painful adjustments for those edged out in the economic competition. Fortunately, many small farms survive and even thrive, along with the neighborly attitude they represent. In the area where I grew up, many farmers still share labor and equipment, just as they do in Colombia and Bangladesh. Such sharing is a sign of hope. But two big concerns are how to meet a global food demand that is expected to almost double over the next 50 years and how to reach the poorest people—about a billion—who have been left behind.

The secret to feeding a nation and the world is not resources, but institutions: the political and economic rules put in place that protect the weak and govern individual incentives for good behavior. A look at the relative well-being of people in the Dominican Republic and Haiti, neighboring countries on the same island, tells us that. The secret is not cultural differences—a look at North Korea and South Korea tells us that. It is not population density—a look at countries with similar densities (such as the Netherlands and Bangladesh) tells us that. Hunger is a problem that can be largely solved with improved, human-made rules and incentives.

As the world becomes increasingly interconnected in commerce, climate, and other ways, international institutions also become important drivers of incentives. Trade agreements, environmental accords, human rights standards, and rules affecting capital flows and technology transfer are just a few examples. Incentives influence good and bad behavior, so they must be implemented with constant vigilance against connivers and quacks.

Some countries, placing priority on reducing poverty and malnutrition, have created incentives to reduce them. Others have not and continue to struggle. Those of us who were born to at least modest means by US standards are fortunate. Our birth has brought us luck, as measured in length and quality of life, and for that we should be thankful. The good news is that when we help others who are less fortunate, we also help ourselves, and for that, too, we should be thankful. People who have the least (such as Julio and Imelda, who donated a pig to help a friend) often give the most—to neighbors and strangers alike. Perhaps they are the ones most thankful for the little things in life—or maybe their intimacy with the problems facing the poor helps them see the need. Certainly, we never truly understand poverty and hunger without standing in the shoes of the poor and hungry.

I have worked in Colombia, Bangladesh, Ghana, and other developing countries long enough to have witnessed much hunger and heartache, but also improvement in even some of the poorest places. That improvement provides hope for the future. When Bangladesh split from Pakistan in 1971, the US media called the country a "basket case," implying it could never feed itself. Its economy would always have to rely on foreign aid to survive. After four decades, Bangladesh is still one of the poorest countries in the world and continues to receive foreign aid. But it now produces roughly enough food to feed itself, and health and nutritional statistics have improved significantly. A sixth of the population remains malnourished, but that statistic is down from more than half at independence. A tripling of rice and vegetable yields has made a difference to the lives of millions of the poor. But increases in crop production, spurred in part by the green revolution, are only part of the story. The spread of microcredit, growth in remittances from workers abroad, free and widespread family planning (which has reduced the fertility rate from more than 6 at independence to 2.3 today), social programs provided by the government and NGOs, and a booming textile industry that has put money in women's pockets have all helped the poor, even if wages are low and safety concerns remain.[75] The reduced fertility rate bodes well for a future "demographic dividend," as more and more of the population will be of working age, saving and investing. Visitors to Dhaka are appalled at the visible poverty, and much remains to be done to improve economic well-being in Bangladesh, but life for many poor people there is better than it was in the past.

In June 2012, I flew to Bogotá and took a bus to the site where we worked in the Peace Corps. Luis Carlos and most of our other Colombian friends no longer live in Pensilvania, and Marj was unable to join me on the trip, but I wanted to visit the farmers with whom we had worked. The US State Department discourages bus travel in rural Colombia, but security has improved and I felt safe enough.[76]

The old seedy downtown bus station in Bogotá has been replaced by a modern facility, and bus comfort has improved significantly. The ride to Pensilvania now only takes about nine hours on mostly paved, though still stomach-churning, roads. Along the way, the bus detoured through Vianí, the sleepy hamlet where I once rode a horse in the plaza during Peace Corp training. It is now a bustling market town. We descended to the Magdalena River, and passed the remnants of Armero, a ghost town where more than 20,000 residents were swallowed up by a river of mud after a 1985 volcanic eruption. We ascended again, and five hours later pulled into Pensilvania.

I went straight to the extension office of the Colombian Coffee Federation and introduced myself to the agronomist in charge. He was cordial and invited me to observe a local radio broadcast he was about to make to farmers. He mentioned on the air that a former Peace Corps volunteer, "Jorge Norton," who had worked with them many years ago with his wife "Margarita," had returned and would welcome visitors in the extension office the next day.

When I walked in at 8:00 AM, Julio and Imelda's daughter Esneda was waiting. She had heard the broadcast and come to greet me. She told me about each of her family members, most of whom are doing well, except for Julio, who is deceased. Her sister Estela now runs a café in town and takes care of Imelda, who is 80. Her brother, Pedro Julio, runs a store and works part-time on the small farm in San Juan where they grew up. Even her brother Fabio, who had polio, is doing well, although he still struggles to walk. He lives in Bogotá where he works and has raised a family of his own.

Esneda took me to see Pedro Julio, and later he took me to their farm to show me his fish pond, goats, garden, and new farmhouse near the road. Over the next several days, he traveled with me to visit several other farms where Marj and I had worked. At one farm, I encountered a woman sitting on the porch knitting with her daughter. Marj had taught her to knit 40 years ago, and as we ate lunch she was proud to show me her work. The woman was the mother of Fidel, the young boy who would occasionally follow me home at supper time. Fidel now lives and works in Bogotá with his wife and children, and they are doing fine, she said. Many other people I visited remembered us from when they were children. Everyone wished Marj had come, and told me how much they appreciated her efforts. I saw a few gardens, but no white rabbits, although people spoke fondly about raising them. As we reminisced— and ate—I realized that gardens and rabbits may have disappeared, but our friendships had not.

Much is the same in Pensilvania as it was 40 years ago, although a lot has changed. The local population is about the same size, but families are smaller and there are many signs of development both in town and on farms. Off-farm jobs have expanded, many farms are now part-time operations, and coffee production is now heavily concentrated on the more progressive farms that have planted varieties resistant to coffee rust disease. Some farms have switched from coffee to dairy production. As incomes have risen, the demand for milk has grown. With electricity now available on farms, dairy producers can cool and preserve their milk until it is marketed. The cows have been improved and adapted to the local environment. I suspect few cows fall off the mountain these days. Most importantly, people on the farms look healthier than they did before.

In town, vehicles and motorcycles have increased substantially, although ironically, so too have horses, as wealth has risen. Meat is now sold from refrigerated cases in shops scattered throughout the town, rather than in a single open-air market. A major recreation center has been established on the edge of town, and of course cell phones and the Internet have replaced the single phone line in town.

One of the most encouraging signs is the improvement in the quantity and quality of education, as evidenced by the computers in rural classrooms, the increased number of children who complete high school, and the establishment of a two-year college that trains students for jobs in the growing lumber and wood industry.

Angela (a granddaughter of Julio and Imelda) hoped to attend a state university in the nearby city of Manizales. She wanted to become an agronomist and give back to her community. A scholarship was made available to her if she studied hard in high school and passed the university admissions test. She worked diligently and last fall received one of the highest scores in the department (state) of Caldas on the exam. She now attends the university. In two generations, the Ospina family has progressed from being extremely poor on a small subsistence farm, with no member having more than a fifth-grade education, to having several members attend high school and run farms and businesses, to having one member on her way to earning a professional degree. The family has gone from hunger to hope and health. Along the way, they have helped many others in their community, and others have helped them.

The Ospinas are not unique. I see their story unfolding daily on farms and in villages around the developing world, even in parts of Africa. It is a story of hard work—Pedro Julio and his wife, Ana, devote long hours to their store and farm so that their children have opportunities to succeed—and sharing along the way. As violence subsides and peace and justice advance, as economic dividends are realized from reduced dependency ratios, as economic and political institutions are improved, and as new technologies increase productivity, the paths to prosperity will multiply, in Colombia and in other countries as well. Problems remain, but there is also progress and hope.

Those of us in developed countries can assist as well. By reaching out to poor people at home and abroad through our institutions and as individuals, we can make a difference in efforts to reduce hunger and malnutrition—to give hope. The day after the tragedy at Virginia Tech, faculty member Nikki Giovanni spoke to the assembled Virginia Tech community and reminded us that "no one deserves tragedy." Certainly the blind beggars in Bangladesh and child soldiers in Africa do not, nor those suffering from malaria and malnutrition around the world. Giovanni suggested that we must embrace our own while reaching out to others "with open hearts and hands. . . . We are better than we think and not quite what we want to be," she said.[77] Progress has been made in alleviating poverty and hunger; there is hope, but we can do more.

Discussion Questions

1. Why do farmers in low-income countries tend to be risk adverse and conservative in their behavior?

2. Why is it almost inevitable that economies experience a transition from a primarily agricultural to a more mixed economy with manufacturing and service sectors?

3. Why is it difficult to truly understand what it means to be poor without walking in the shoes of a poor person?

4. Many college students are poor. In what ways might their mental outlooks be different from poor people in low-income countries?

5. In what ways do Peace Corps volunteers gain more than they give?

6. Why is sharing so prevalent in developing countries?

7. Is rapid population growth a problem in developing countries? Why or why not?

8. Why are institutional changes as important, or more important, than improved technologies for reducing poverty and hunger problems in the world?

9. Why did George and Marj accept food from poor, hungry people? What would you do if offered food by a poor person?

10. What is the hungry season? Why does it occur? How do people cope with it?

11. George and Marj lived on a low salary in the Peace Corps, but they were among the richest people in their town. Why might it be difficult to live as one of the richest people in town?

12. Ecuador is a poorer country than the United States. How has it been able to make at least a minimum level of medical care available to everyone?

13. Poverty differs by degree. Why do some poor people despise others who are poorer than themselves, while other poor people graciously share with their poorer neighbors?

14. How might cultural factors hinder agricultural and economic development and how might they help development?

15. The importance of institutional change was stressed in the book, but why are improved agricultural technologies also important for agricultural development?

16. Why is education so important for agricultural development and poverty reduction?

17. How can international trade help reduce hunger? How can it make it worse?

18. Why might both the presence of local foods and international trade improve diets?

19. Why might foreign aid help reduce poverty and hunger, and what are the concerns with foreign aid?

20. Why is agricultural development so important for overall economic development?

21. Poverty and hunger were described in the book, but in what sense is there hope for reducing them?

22. Why are having children so important to families in the least developed countries?

23. Why do many farms grow larger as economic development occurs, while so many farms keep operating as small, part-time farms?

Notes

Prologue

[1] The origin of the town's American looking name of Pensilvania is a mystery. One person told me that someone from Pennsylvania had passed by when the town was founded in the mid-1800s and had influenced the choice of name, but that person had had a few drinks when he told me. An explanation does not appear in the documented town history.

[2] History is full of examples of such unequal growth and subsequent conflict, from the Roman empire to the Mayan empire to the recent conflicts in several African nations.

[3] In the terminology of Daron Acemoglu and James Robinson, *Why Nations Fail: The Origins of Power, Prosperity, and Poverty* (New York: Crown Business, 2012), the political and economic institutions in Ireland in the early and mid-1800s were "extractive" rather than "inclusive."

[4] For a more detailed description of rural Ireland at the time, see: http://www.historyplace.com/worldhistory/famine/index.html. By the time a major outbreak of late blight (*Phytophthora infestans*) from 1845 to 1852 had finished devastating the potato crop, the Irish population of eight million inhabitants had dropped to six million.

[5] Patrick may have thought his family name was a bit too close to naughty, as he changed it to the Anglicized spelling, Norton. A more likely explanation is that at the time, the Irish were being discriminated against, so Patrick Anglicized his name for easier assimilation into American society.

[6] The original Naughton farm where Patrick grew up was passed to his brother, John, when their father died. The farm was only marginally productive, and after the house burned in the 1890s, the farm was eventually sold, and its new owner allowed it to go back to forest. The fates of John and Patrick's other brothers and sisters are described in Mary Clarke Norton, "The Norton and Will Families of Hillsboro," Oneida County, New York, mimeo, May 1962. Hillsboro is four miles west of Camden, New York.

[7] In 1790, 90 percent of US labor was in agriculture. When Patrick Naughton arrived from Ireland in 1838, it had dropped to about 70 percent. The total farm labor force grew until it reached about 32 million workers in 1910, in part due to a high rate of natural population growth among farm families and immigration, but by then farm labor had dropped to about 30 percent of the total US workforce. At the end of World War II it was 16 percent, by 1970 less than 5 percent, and in the 1990s it bottomed out at around 2 percent of the workforce. Percentage wise, the biggest squeeze on US farms began after World War II and lasted until the early 1990s. During that period, the number of farms dropped by two-thirds. Carolyn Dimitri, Anne Effland, and Neilson Conklin, *The 20th Century Transformation of U.S. Agriculture and Farm Policy* (EIB-3) (Washington, DC: Economic Research Service, US Depart-

ment of Agriculture, 2005); Growing a Nation: The Story of American Agriculture, Debra Spielmaker, Project Director (http://www.agclassroom.org), accessed February 2013.

[8] During this development process, increased agricultural productivity provides food, labor, and a market for overall economic development. For more details, see: George W. Norton, Jeffrey Alwang, and William A. Masters, *Economics of Agricultural Development: World Food Systems and Resource Use* (New York: Routledge, 2010, pp. 90–92).

[9] The United States invested in the land grant system that supported at least one college of agriculture and engineering in each state. It also invested in experiment stations that focused on agricultural research. Later it supported an agricultural extension system that helped to diffuse the new technologies to farmers.

[10] Agro-tourism involves agricultural activities that bring visitors onto a farm or ranch. It includes activities such as buying produce direct at a farm stand, picking fruit on a farm, feeding animals, or staying at a bed and breakfast on a farm.

[11] Professors David Thurston and Gordon Cummings at Cornell.

[12] Dressed in filthy clothes, *gamines* still wander the streets of Bogotá today. A few of the kids are orphans or abandoned, but most of them leave their homes in the slums of their own accord. They abandon their families to live on the streets because of mistreatment, hunger, or neglect. Often regarded as the dregs of society, many of these kids had the smarts to escape an abusive situation as soon as they could. On the street, they survive on their own wits and typically become part of a communal culture. Life is harsh, but the street groups these children join become substitute families and provide them with a basic level of social interaction and security. The younger kids beg and pick pockets, while older ones steal, run drugs, or sell items such as lottery tickets, knickknacks, or informal services. They value their independence, but often find it difficult at adolescence to adjust to adult society. They are tough, wily, and frequently embark on a life of crime. Additional insights into the lives of Colombian *gamines* are found in: Lewis Aptekar, "Are Colombian Street Children Neglected? The Contributions of Ethnographic and Ethno-Historical Approaches to the Study of Children," *Anthropology and Education Quarterly*, 22(4) (December 1991): 326–349.

[13] Among other items, the dashboard was decorated with small statues of St. Christopher and other patron saints of safe travel.

[14] Peace Corps volunteer Michael Kotzian died in Colombia on March 17, 1969.

[15] The origin of the name Filadelfia is unclear, although it is a Spanish translation of the Greek word for brotherly love.

[16] Information in this paragraph was obtained from: Kenneth Carley, *The Dakota War of 1862* (St. Paul: Minnesota Historical Society Press, 1976).

[17] The tribe recently substituted the Dakota word "Oyate" for "Sioux Tribe" in their tribal name. It means "people."

[18] Support was provided by Winrock International and Northwest Area Foundation.

[19] Seven clans make up the Sisseton Wahpeton Oyate.

Chapter 1   Poverty Is Personal

[20] Economist Adam Smith first postulated in the eighteenth century that the greatest benefit to society is brought about by individuals acting freely in a competitive marketplace in pursuit of their own self-interest. This self-regulating nature of markets has been used to justify a laissez-faire economic philosophy, and has prompted debate about the appropriate role of public policies that interfere with markets.

21 The foundation that was established in the subsequent months is located in La Lima, Honduras, and is called FHIA (Honduran Agricultural Research Foundation).

22 Kamala Markandaya, *Nectar in a Sieve* (New York: Signet Classics, 2002).

23 World Health Organization, Statistical Information System (http://www.who.int/whosis/whostat/2009/en/index.html).

24 Seth Tillman, "The Ability to Forget," *The Peace Corps Reader,* published for the Peace Corps by Quadrangle Books, 1967, p. 75.

Chapter 2   Fertility Fears

25 For a revealing look at medical conditions and child birth in a Malian rural village, read: Kris Holloway, *Monique and the Mango Rains: Two Years with a Midwife in Mali* (Long Grove, IL: Waveland Press, 2007).

26 Paul Ehrlich, *The Population Bomb* (New York: Ballantine Books, 1968).

Chapter 3   Hunger Hurts

27 Kamala Markandaya, *Nectar in a Sieve* (New York: Signet Classics, 2002), p. 91.

28 For more information, see: Michelle Adato and John Hoddinott, "Conditional Cash Transfer Programs," *2020 Focus Brief on the World's Poor and Hungry People* (Washington, DC: International Food Policy Research Institute, October 2007).

29 Brazil, for example, slashed stunting rates in its impoverished northeast from 34 percent in 1986 to 6 percent in 2006 (Richard Kerr, "The Greenhouse is Making the Water-Poor Even Poorer," *Science*, Vol. 336, April 27, 2012, p. 402).

Chapter 4   Feeling Sick

30 Freddy Krueger was a character in the *Nightmare on Elm Street* series of horror movies. I have to admit I have never had the urge to see one of the movies in this series, but if you have seen the ads you get the picture.

31 Multiple sclerosis (MS) is a disease that affects the central nervous system. Symptoms may be mild, such as numbness in the limbs, or severe, such as paralysis or loss of vision. The progress, severity, and specific symptoms of MS differ from person to person. With MS, the body's own immune system attacks myelin, the substance that surrounds and protects the nerve fibers in the central nervous system. Nerve impulses traveling to and from the brain and spinal cord are distorted or interrupted, producing a variety of symptoms such as fatigue, numbness, difficulty with walking, balance, and coordination, pain, incontinence, vision problems, spasticity, cognitive issues, and depression. People with MS can experience different paths of the disease. Since no two people have exactly the same experience with MS, the disease may look very different from one person to another. More information about MS can be found at http://www.nationalmssociety.org/index.aspx.

32 Recently an officer of Alpha Gamma Rho fraternity at Virginia Tech came up to me and asked if I still traveled a lot overseas. I said yes. He said he had heard that Marj has MS and that she should call their house if she ever needs help when I am gone. The fraternity brothers would come over to assist her. It is nice to know that someone has got your back.

[33] United Nations Development Program, Human Development Report 2007–2008 (New York: Palgrave, 2007) (http://hdr.undp.org/en/media/HDR_20072008_EN_Complete.pdf).

[34] Gijs Walraven, *Health and Poverty: Global Health Care Problems and Solutions* (London: Earthscan, 2011).

Chapter 5   Fragile Fields

[35] Most insecticides are neurotoxins. A high enough dosage will affect the human nervous system.

[36] Monsoons in South Asia are caused by seasonal winds off the Bay of Bengal and the Arabian Sea that bring heavy rainfall to places like India and Bangladesh during the summer months. Typhoons are essentially Asian hurricanes.

[37] Richard Kerr, "The Greenhouse is Making the Water-Poor Even Poorer," *Science*, Vol. 336, April 27, 2012, p. 405.

[38] Natural Resources Conservation Service, *2007 National Resources Inventory* (Washington, DC: US Department of Agriculture, April 2010).

[39] See Edward Barbier, *A Global Green New Deal* (New York: Cambridge University Press, 2010).

Chapter 6   It Takes a Farm

[40] Kamala Markandaya, *Nectar in a Sieve* (New York: Signet Classics, 2002).

[41] His coffee had "mancha de hierro" (*Cerocospora coffeicola*), which is a common coffee disease, but not as serious as others such as "ojo de gallo" (*Mycena citicolor*) and "la roya" (*Hemileia vastatrix*). It attacks the leaves and the coffee berries and can be identified by small circular lesions on the leaves that are gray in the center and surrounded by a darker ring and then a yellowish ring.

[42] The disease is bacterial wilt (*Ralstonia solanacearum*).

[43] That trick also works in houses. I was speaking with a woman in Colombia one day when we noticed a large number of crawling ants on her porch. She scooped up a few ashes from her stove and sprinkled them on the floor. Within a couple of minutes there was not an ant to be found.

[44] We also worried about risks such as drought and fire. One summer, a shed that contained our hay-baler and several tons of stored hay caught fire and burned. My father tried desperately to save the baler, but the flames were just too hot. It was the only time as a child that I ever saw my father cry.

[45] See: Barbara Kingsolver, *Animal, Vegetable, Miracle* (New York: Harper Collins, 2007).

Chapter 8   Building Bridges

[46] October 16, 1972.

Chapter 9   Seeds of Hope

[47] These centers are collectively known as the Consultative Group for International Agricultural Research (CGIAR), or the Future Harvest Centers. The first two centers were the

International Rice Research Institute located in the Philippines and the International Maize and Wheat Improvement Center in Mexico.

48 Some of which are included in: Julian M. Alston, George W. Norton, and Philip G. Pardey, *Science under Scarcity: Principles and Practice for Agricultural Research Evaluation and Priority Setting* (Ithaca, NY: Cornell University Press, 1995).

49 Evaluation studies have estimated more than a half billion dollars in benefits from the use of IPM CRSP technologies and information by farmers. The formal name of the IPM CRSP was recently changed by USAID to IPM Innovation Lab (IPM IL).

50 See: James R. Stevenson, Nelson Villoria, Derek Byerlee, Timothy Kelley, and Mywish Maredia, "Green revolution research saved an estimated 18 to 27 million hectares from being brought into agricultural production," *Proceedings of the National Academy of Sciences*, May 21, 2013, 110(21), p. 8367, who estimate that up to 43 million hectares (103 million acres) would have been needed to achieve the food consumption that occurred.

51 Including a potentially devastating strain called UG99, which first began to cause losses in East Africa in 1998 before spreading through the Horn of Africa and into the Middle East.

Chapter 10   It Takes a School

52 UN Human Development Report 2010 and 2011 HDI Education Indicator, accessed September 2011 at: http://hdr.undp.org/en/statistics/data/2011/.

53 Dean Karlan and Jacob Appel, *More Than Good Intentions: How a New Economics is Helping to Solve Global Poverty* (New York: Dutton, 2011), p. 206.

Chapter 11   Policy Pitfalls

54 A similar problem can occur with a cow that is left too long to feed in a field of fresh alfalfa or clover.

55 I learned the details of this story from the extension agent.

56 The bank was started by Mohammed Yunus, who later won the Nobel Prize for the innovative microcredit effort.

Chapter 12   Going Global, Buying Local

57 Several major international clothing companies, including H&M, the largest purchaser of garments from Bangladesh, have signed contracts with the Bangladesh government, agreeing to make independent safety inspections, make reports on factory conditions public, cover the cost of repairs, and stop doing business with any factory that refuses to make necessary safety adjustments. However, some companies have not signed such contracts, and one company, Disney, chose to stop doing business with garment factories in Bangladesh.

58 EUREPGAP stands for Euro-Retailer Produce Working Group Good Agricultural Practices and was established in 1997. To reflect its global reach, it changed its name in 2007 to GlobalGAP.

Chapter 13   Cultural Clues

[59] For a revealing look at the rights of women in Malian village culture, read: Kris Hollo-way, *Monique and the Mango Rains: Two Years with a Midwife in Mali* (Long Grove, IL: Waveland Press, 2007).

Chapter 14   Avoiding Violence

[60] Daniel Wilkerson, *Silence on the Mountain: Stories of Terror, Betrayal, and Forgetting in Guatemala* (Durham: Duke University Press, 2004), p. 212. Details in this paragraph and the preceding one are taken from this book.

[61] Robert Dix, *Colombia: The Political Dimensions of Change* (New Haven: Yale University Press, 1967).

[62] Marj and I once witnessed political violence on a small scale in Pensilvania. We were sitting in a cafe talking to a friend the day after a Colombian national election. The day before, all the bars in town had been closed as a precaution against mixing politics with booze. Of course the men in town relieved their thirst the following day by doubling down on drinks. In the cafe, we noticed a few guys who had a little too much local brew. Suddenly, a fight broke out at a nearby table. One man smashed a chair over the head of another. Others joined in and soon nearly everyone in the room was fighting. Participants seemed to know who favored which political party and fists were flying fast and furious. We ducked out the door, but the fight quickly spread to the plaza and surrounding streets where others joined in. The massive brawl would have seemed almost comical if it were not so serious. Not having a stake in the fight, we hurried back to our apartment two blocks away. From the doorway we could see police struggling to control the mob, but they were soon overwhelmed. Ironically, they called Manzanares for back-up, and that police force arrived in the back of a dump truck. The truck drove up and down the street arresting people and enforced a curfew until things calmed down. Fortunately no one was seriously hurt that day.

[63] The Centro Internacional de Agricultura Tropical (CIAT) is located in Colombia, but has a global mandate. It is funded by many public and private donors and is part of the Consultative Group for International Agricultural Research (CGIAR) system.

[64] See: Douglas North, John J. Wallis, and Barry Weingast, *Violence and Social Orders: A Conceptual Framework for Interpreting Recorded Human History* (New York: Cambridge University Press, 2009).

[65] For evidence see: Maarten J. Voors, Eleonora E. M. Nillesen, Philip Verwimp, Erwin H. Bulte, Robert Lensink, and Daan P. Van Soest, "Violence and Conflict: A Field Experiment in Burundi," *American Economic Review*, 102(2) (2012): 941–964.

Chapter 15   Doing Well by Giving Goods

[66] Mesfin Bezuneh, Brady J. Deaton, and George W. Norton, "Food Aid Impacts in Rural Kenya," *American Journal of Agricultural Economics*, 70(1) (February 1988): 181–191.

[67] The Nepal Smallholder Irrigation Market Initiative (SIMI) was supported by several donors including Swiss, German, Dutch, and US aid agencies, as well as FAO, the World Bank, and others. It was implemented by Winrock International and International Development Enterprises in partnership with numerous Nepalese government agencies and nongovernmental organizations.

[68] See Per Pinstrup-Andersen and Derrill D. Watson II, *Food Policy for Developing Countries* (Ithaca, NY: Cornell University Press, 2011), p. 212.

[69] Comment by Dr. Toledo at a Peace Corps fiftieth anniversary event, Washington DC, September 24, 2011, as reported by Erica Burman, "Captivating Conversations: Global Leaders Reflect on Peace Corps Impact and Legacy," *Worldview*, Winter 2011–12, p. 12.

[70] For a description of farm life for a young woman in India, read Kamala Markandaya, *Nectar in a Sieve* (New York: Signet Classics, 2002).

[71] In the months after the shootings in April 2007, the Virginia Tech community was the recipient of many kind gestures. The Dave Matthews Band, John Mayer, and Phil Vassar performed an uplifting free concert on campus. Donations poured into the fund for the victims of the tragedy from around the nation and world. Students and faculty, struck by the stories of kindness and accomplishments of those who perished and wanting to move forward while honoring their memories, expanded their own efforts to reach out to others in the community. Students made the Relay for Life at Virginia Tech one of the largest on any college campus. They built footbridges in developing countries such as Haiti to help people access medical care, schools, and economic opportunities. They assisted and engaged the elderly in their homes. Most importantly, they showed heightened awareness of the emotional needs of those around them every day. They turned grief into good.

[72] William Easterley, *The Elusive Quest for Growth: Economists Adventures and Misadventures in the Tropics* (Cambridge, MA: MIT Press, 2001).

[73] Paul Collier, *The Bottom Billion: Why the Poorest Countries Are Failing and What Can Be Done about It* (New York: Oxford Press, 2007).

[74] George W. Norton, Jeffrey Alwang, and William A. Masters, *Economics of Agricultural Development: World Food Systems and Resource Use* (New York: Routledge, 2010), chapter 12.

Chapter 16   Finding Hope

[75] Garment workers in Bangladesh make about $2.00 per day, and safety deficiencies in apparel factories have cost many workers their lives (as described in chapter 12). Safety problems, such as the lack of or locked fire exits, in the factory where 117 people died in November 2012, were similar to problems at the infamous Triangle apparel factory in New York City, where 149 people were killed in 1911. The Triangle fire led to new worker safety regulations in the United States. In Bangladesh as well, the major fire and the horrific building collapse may provide the impetus to improve worker safety. The threat of losing international clothing buyers may cause the country to finally get serious about safety. Bangladesh exports $20 billion worth of apparel, which represents 80 percent of the country's exports. Concern for their reputation and sales recently led more than 30 companies that source garments from Bangladesh to sign an agreement to conduct independent factory inspections and provide support to implement safety measures.

[76] As I write this book, the government of Colombia is negotiating a peace plan with the main guerilla group, and kidnappings are less frequent than they were.

[77] For the complete remarks of Nikki Giovanni and others that day, see: http://www.remembrance.vt.edu/2007/archive/giovanni_transcript.html. The Virginia Tech community has gradually healed since April 16, 2007, a process facilitated by reaching out to others in need.

Suggested Readings

Daron Acemoglu and James A. Robinson, *Why Nations Fail: The Origins of Power, Prosperity, and Poverty* (New York: Crown Business, 2012), 529 pages. A wealth of historical examples of why institutions matter for development.

Katherine Boo, *Behind the Beautiful Forevers: Life, Death, and Hope in a Mumbai Undercity* (New York: Random House, 2012), 256 pages. A well-documented story of life in a shanty town in India.

Paul Collier, *The Bottom Billion: Why the Poorest Countries Are Failing and What Can Be Done About It* (New York: Oxford Press, 2007), 209 pages. An insightful book that describes how "conflict traps" can keep countries poor.

Tony D'Souza, *Whiteman* (Orlando, FL: Harcourt Press, 2006), 379 pages. An exceptional tale of West African village life though the eyes of a fictional American relief worker.

William Easterly, *The Elusive Quest for Growth: Economists' Adventures and Misadventures in the Tropics* (Cambridge, MA: MIT Press, 2001), 342 pages. A readable reminder of why incentives matter to the success of development policies.

Kris Holloway, *Monique and the Mango Rains: Two Years with a Midwife in Mali* (Long Grove, IL: Waveland Press, 2007), 215 pages. A poignant book about life in an impoverished village in Mali.

Dean Karlan and Jacob Appel, *More Than Good Intentions: How a New Economics Is Helping to Solve Global Poverty* (New York: Dutton, 2011), 308 pages. A well-written account of how proposed changes in credit, health care, education, and other development initiatives can be carefully evaluated to improve the well-being of the poor.

Tracy Kidder, *Mountains beyond Mountains: The Quest of Dr. Paul Farmer, the Man Who Would Cure the World* (New York: Random House, 2004), 317 pages. An enlightening story of one doctor's quest to improve health in rural Haiti and beyond.

Kamala Markandaya, *Nectar in a Sieve* (New York: Signet Classics, 2002), 190 pages. This classic best seller provides a moving portrait of the life of a woman married to a poor farmer in India.

George W. Norton, Jeffrey Alwang, and William A. Masters, *Economics of Agricultural Development: World Food Systems and Resource Use,* 2nd edition (New York: Routledge, 2010), 465 pages. Third edition forthcoming in 2014. Textbook coverage of topics addressed in this book.

J. Edward Taylor, *Essentials of Development Economics* (Berkeley, CA: Rebel Text, 2013), 290 pages. A readable textbook on development economics.

Mortiz Thomsen, *Living Poor: A Peace Corps Chronicle* (Seattle: University of Washington Press, 1990), 280 pages. The most-read Peace Corps memoir ever.

Roger Thurow, *The Last Hunger Season: A Year in an African Farm Community on the Brink of Change* (New York: Public Affairs, 2013), 328 pages. A clearly written chronicle of the lives and work of limited-resource farmers in western Kenya and the steps they take to feed their families and improve their situations.

Eve Brown-Waite, *First Comes Love, Then Comes Malaria* (New York: Broadway Books, 2009), 306 pages. The author transforms herself in the developing world, while laughing away the stresses.

Gijs Walraven, *Health and Poverty: Global Health Care Problems and Solutions* (London: Earthscan, 2011), 192 pages. Practical solutions to health care problems in developing countries.

Websites

Hunger and Hope (http://hungerandhope.agecon.vt.edu)—website for this book.

International Food Policy Research Institute (http://www.ifpri.cgiar.org)—a wealth of carefully researched information on agricultural development.

Economics of Agricultural Development (http://ecagdev.agecon.vt.edu)—textbook site.

Giving Thanks

A great many people influenced the content and preparation of this book, beginning with my Wooster School English teacher, Korb Eynon. I entered his class thinking I knew how to write, and quickly discovered I knew little, which was worth a lot. Thank you, Korb. I also thank Wooster teacher Joe Grover for believing in me.

I thank one of my undergraduate professors at Cornell, John Mellor, for helping stimulate my interest in agricultural development. I thank Bob Thompson for drawing Dr. Mellor to my attention, and for being such a helpful and inspirational brother-in-law all these years. Your advice and wisdom are highly valued. While on the subject of in-laws, let me thank Robert and Esther Thompson for being so supportive and trusting of me with your daughter. I know it was not easy to see her dragged off to Colombia for two years where she bore your grandchild in an unknown land, or to see her deserted for weeks at a time since then while I wander the world. On a similar note, thanks to brothers- and sisters-in-law Bob, Marilyn, Bill, Ruth, Mark, Donald, David, Stephen, and Marcia for welcoming me so warmly into your fabulous family.

I thank my PhD advisor, Bill Easter, and my master's advisors, Bill Easter and Martin Abel, who were everything a student could ask for. I also thank my grad school office mate, John Spriggs, from whom I learned an immeasurable amount of economics, which may not be much since it cannot be measured, but I really am grateful, John. I thank several current and former Virginia Tech faculty colleagues for research collaboration and advice: Jeff Alwang, Brady Deaton, Darrell Bosch, David Orden, Dixie Reaves, S. K. De Datta, Brad Mills, Dan Taylor, Jason Grant, Everett Peterson, Mike Ellerbrock, Randy Kramer, Joe Coffey, Sandra Batie, Leonard Shabman, Jonathan Nevitt, Muni Muniappan, and Short

Heinrichs. I especially thank Jeff, who has worked with me more than anyone else for the past 20 years, for being such a great mentor, loyal friend, and caring person. I thank S. K. for being steadfastly supportive, even when I do dumb things.

I thank numerous professional colleagues at other institutions in the United States and abroad for research collaboration and advice, including Ed Rajotte, Sally Miller, Bob Hedlund, Scott Swinton, K. L. Heong, Phil Pardey, Julian Alston, Stan Wood, Will Masters, Jerry Shively, Burt Sundquist, Randy Barker, Mesfin Bezuneh, Wally Huffman, Paul Heisey, Dave Schimmelpfennig, Luke Colavito, Gershon Feder, Bob Herdt, Derek Byerlee, Howard Elliott, Carl Pray, Shenggen Fan, Rezaul Karim, Victor Barrera, Pedro Pablo Pena, Serge Francisco, Patricio Espinosa, Julio Palomino, Santi Obien, Maripaz Perez, Awere Dankyi, Jackie Bonabana, Majd Sayedissa, Ruben Echevarria, Carlos Pomareda, Victor Ganoza, Agnes Rola, Emmanuel Okogbonin, Cheidozie Egesi, Abdelbgi Ismail, Modan Dey, and Jose Falck-Zepeda. It is hard to thank Ed enough for more than 30 years of collaboration and friendship, always making sure we saw the humor in every bit of craziness we encountered, and Sally for playing the straight woman to Ed and for her friendship. I learned a lot about bugs and plant diseases from you both, although not as much as I pretend to know. Thank you, Darrell, for being there for Marj when needed, and for celebrating and commiserating with me over the Twins and Vikings.

I thank Luis Carlos Giraldo for teaching us the ropes and being such a good friend during our Peace Corps days. Also from Peace Corps, I thank Sue Garner, Ginny de Victoria, Salomon Sanchez, Joe and Meredith Kwiatkowski, Jim and Lynn Cornell, John Miles, Cindy Ross, Mauricio Mariño, and our many friends in Pensilvania. I thank the Peace Corps program itself for giving me the most eye-opening experience of my life, a new way of looking at the world, and an understanding of some of the intricacies and challenges of another culture and human condition.

I express my appreciation to Jerry Flute, Ed Red Owl, and Mike Selvage for welcoming me and teaching me about the life, culture, and concerns of the Sisseton Wahpeton Oyate, and to Jim Hammill, now retired from the Federal Reserve Bank in Minneapolis, for putting your trust in a young graduate student. Thank you, Jim and Lyn Gray, for dedicating so much of your life to improving the lot of people in Liberia.

A professor is always indebted to his or her graduate students because they do so much of the thinking and work! Thank you Blair, Fred,

Edward, Cathy, Steve, Eddie, Maurice, Krishna Napit, Kalyani, Rich, Jaime, Pedro, Sang, Julio, Brady, Jeff, Jessica Tjornhom, Nicos, Nihal, Ari, Gladys, Shukla, Binshang, Paul, Ebere, Leah Cuyno, Lefter, Jim, Tak, Tom, Jason Beddow, Cesar, Guy, Jason Bergtold, Luis, Maria, Carolina, Mishra, Sibusiso, Jacob, Genti, Melanie, Atanu, Jessica Bayer, Yan, Vida, Aurora, Daisy, Jason Maupin, Nderim, Pricilla, Catherine, Tatjana, Theo, Adam, Zalalem, Leah Harris, Abby, Gertrude, Vanessa, Krishna Parajuli, Will, Jianfeng, Kiruthika, Barry, Kate, Jarrad, Evan, and Alda. Thanks to each of you for your dedication, for being friends and colleagues, and for putting up with my quirks.

Several people assisted immensely with their reviews of previous drafts of this book, including Gyorgyi Voros, who provided extensive comments on two drafts, Su Clauson-Wicker, Frieda Bostian, Karen Glass, Nadia Tuck, Michelle Klassen, Sally Brady, John Norton, Richard Norton, Jeff Alwang, Mike Ellerbrock, Ed Rajotte, Jessica Boatwright, Liz Philips, Stephanie Myrick, Catherine Goggins, Jane Abraham, Bill Easter, and Marj. Your comments and suggestions are much appreciated. I also thank my editors at Waveland Press, Don Rosso and Diane Evans, for your support and attention to detail. I thank the students in AAEC 3204, 5154, and 5174 for being attentive (most of you), for helping sharpen my own thoughts on agricultural development through your questions, and for listening to my stories and dumb jokes. Thanks to Sandy Yore at Martin Travel for figuring out the best way to get my students and me to all the obscure places we have been over the past many years—I didn't really mean that quip in the prologue about my travel agent.

I thank my brothers, Richard and Charles, for being fantastic friends and for marrying such wonderful wives. You keep me grounded. I would also like to acknowledge the contribution of my late grandmother, Mary Clarke Norton, in researching the Norton history summarized in the prologue, and of my late mother, Janet Williamson Norton, for saving all my letters and e-mails that supported stories in this book.

Finally and foremost, I thank Marj and John for being the most loving wife and son a person could ever have. I appreciate all the sacrifices you have made over the years so I could continue to travel. Marj, I can tell from the look in your eyes when I leave for the airport and from the enthusiasm of your hugs when I return, it has not been easy.

About the Author

George Norton is a professor of agricultural and applied economics at Virginia Tech and lives with his wife Marjorie in Blacksburg, Virginia. George and Marjorie served as Peace Corps volunteers in Colombia from 1971 to 1973.

(Photo by John Norton)